Settling Hebron

THE ETHNOGRAPHY OF POLITICAL VIOLENCE

Tobias Kelly, Series Editor

A complete list of books in the series is available from the publisher.

SETTLING HEBRON

Jewish Fundamentalism
in a Palestinian City

Tamara Neuman

PENN

UNIVERSITY OF PENNSYLVANIA PRESS

PHILADELPHIA

Copyright © 2018 University of Pennsylvania Press

All rights reserved. Except for brief quotations used for purposes of review or scholarly citation, none of this book may be reproduced in any form by any means without written permission from the publisher.

Published by
University of Pennsylvania Press
Philadelphia, Pennsylvania 19104-4112
www.upenn.edu/pennpress

Printed in the United States of America on acid-free paper
1 3 5 7 9 10 8 6 4 2

Library of Congress Cataloging-in-Publication Data
Names: Neuman, Tamara, author.
Title: Settling Hebron : Jewish fundamentalism in a Palestinian city / Tamara Neuman.
Other titles: Ethnography of political violence.
Description: 1st edition | Philadelphia : University of Pennsylvania Press, [2018] | Series: Ethnography of political violence | Includes bibliographical references and index.
Identifiers: LCCN 2017049078 | ISBN 9780812249958 (hardcover : alk. paper)
Subjects: LCSH: Jews—West Bank—Hebron. | Land settlement—West Bank—Hebron. | Jewish fundamentalism—West Bank—Hebron. | Hebron—Ethnic relations. | Hebron—in Judaism.
Classification: LCC DS110.H4 N48 2018 | DDC 956.94/2—dc23
LC record available at https://lccn.loc.gov/2017049078

Contents

Note on Transliteration, Translation, and Terms	vii
Introduction	1
Chapter 1. Orientations	27
Chapter 2. Between Legality and Illegality	50
Chapter 3. Motherhood and Property Takeover	70
Chapter 4. Spaces of the Everyday	96
Chapter 5. Religious Violence	122
Chapter 6. Lost Tribes and the Quest for Origins	151
Conclusion: Unsettling Settlers	179
Notes	193
References	211
Index	225
Acknowledgments	239

Note on Transliteration, Translation, and Terms

I have translated the passages in the book except those excerpted from sources translated into English as included in the list of references. The transliterations in the book follow the Library of Congress rules for Hebrew transliteration, substituting the consonants accordingly:

א	ʾ
בּ	b
ב	v
ג	g
ד	d
ה	h
ו	v
ז	z
ח	ḥ
ט	ṭ
י	y (only if consonant)
כּ	k
כ	kh
ל	l
מ (and final ם)	m
נ (and final ן)	n
ס	s
ע	ʿ
פּ	p
פ (and final ף)	f
צ (and final ץ)	ts
ק	ḳ

ר	r
שׁ	sh
שׂ	ś
ת	t

The definite article (ha-, he-), the conjunction (u-, va-, ve-), and certain prepositions (e.g., b, k, l, m) have been separated by hyphens from the words to which they are prefixed. For proper names and Hebrew terms as well as Arabic terms in the text that are commonly used in English speech or writing, I have preserved conventional spellings.

Throughout the text I have used the term "*ideological* settler" (and at times "religious fundamentalist") where others have chosen to use the term "Modern Orthodox" or "National Religious" (*dati-leumi*) to refer to this sector in the Israeli context. These groups have historically synthesized far right or ultranationalist and religious elements to shape a distinct kind of Jewish observance. Modern Orthodoxy, however, includes religious communities that do not live in ideological settlements (and not all in the Modern Orthodox camp would be sympathetic to ideological settler views). For this reason, I refer to the far-right spectrum of religious settlers living in settlements like Kiryat Arba and the Jewish Quarter in Hebron as "*ideological* settlers" and distinguish them from those moving into settlements for economic or "quality of life" reasons alone.

The names of key informants in this ethnography as well as other identifying details have been changed to preserve anonymity. The exceptions to these are as follows: those who are already well known public figures or religious right activists within ideological settler circles, or those who have published various memoirs or studies. I have also retained the names of several former Kiryat Arba residents who are no longer living, or whose lives others have written about in publications that memorialize their thoughts and deeds.

Figure 1. Israeli settlements and outposts in the West Bank by population. Americans for Peace Now.

Introduction

Early on in my fieldwork, Rivka Ashkenazi, an elderly Parisian resident of the West Bank settlement of Kiryat Arba took me to see the Tomb of the Patriarchs in the city of Hebron. As we walked before the monumental structure, she explained that this was the burial site of Judaism's most important matriarchs and patriarchs, and that few nations exist today that know exactly where their ancestors lie. It was in this sense, she continued, that the Jewish people were distinct. Approaching the massive outer walls of this seventh-century site, I took in the vast military panorama encircling the area—observation towers, camouflage netting, barbed wire, steel fencing, metal detectors, and checkpoints. Two towering square minarets, rising up from the diagonal corners of the site's rectangular outer wall, stood as staunch witnesses to the site's Islamic character. As I further observed the scene, Rivka recounted the suggestion by medieval Jewish philosophers that the area stood at the entrance to the Garden of Eden. The dissonance of seeing this heavily militarized zone while hearing her claim that we were standing before Eden remains etched in my mind to this day. Her projection of a biblical utopia onto an elaborate latticework of militarism was telling. Realities, to be sure, can be parsed in myriad ways, but it seemed impossible not to notice the deadening effects of the many soldiers deployed throughout a Palestinian urban area. Rivka's Eden was part of a claim to an exclusive site of Jewish origin, underwritten by a sense of permanent belonging. This claim has great existential as well as political ramifications. When seen through the lens of religious settlement, much of the conflict it has fueled comes about by creating resolute ties to recreated Jewish sites in Palestinian areas and making changes in the landscape to affirm their self-evident biblical link to the past.

On Seeing and Believing

It might seem easy to dismiss Rivka's assertions as an illusion. Yet in remaking and residing in sacred places such as these, Jewish settlers establish a putative sense of the real, which arises from the very materiality of the scene. Being able to see in this particular way, to look beyond the presence of actual Palestinian lives and be invested in Jewish origins alone, comes from the ability to bound off discordant elements of an ideological vision as "alien" or as falling outside an arena of concern. Yet in fact Rivka was confronted by an array of conditions that might in other circumstances have disrupted her religious vision. There was no mistaking, for instance, the crumbling state of many uninhabited Palestinian buildings that had fallen into disrepair or the tension palpable in this volatile and conflict-ridden zone. Rivka's principal focus, however, was on reclaimed Jewish spaces and origins. Her vision was enmeshed in a biblical sense of place and shaped by a mystically rooted experience of self quite unknown in other times and contexts of Jewish observance.

In this book, I analyze the discourses, values, and practices through which ideological settlers remake Palestinian Hebron as a site of Jewish origins in the context of the militarily occupied West Bank to create a rationale for permanently controlling territory in these areas.[1] Rivka's way of seeing has many resonances with those of earlier immigrant settlers or labor Zionists in Palestine by virtue of erasing the Palestinian presence she encounters, but it also reveals a number of features that distinguish her sensibility as unique. By addressing this distinct iteration of settlement, I aim to give ideological settlement the focused attention that it warrants, while situating it within a wider set of social transformations and ruptures that give it resonance within the context of Zionism and Israeli nation building. There are several conjoined processes that have worked together to propel Jewish immigration to Palestine over the course of a century. While settlement has figured distinctly at each stage, it has not taken the syncretic form of ideological or devoutly religious settlement that we find in the present, as it manifests itself in Kiryat Arba and Hebron in particular, as well as in other fundamentalist Jewish settlements.

Located in the West Bank and established on confiscated Palestinian land in 1971, Kiryat Arba consists of approximately seven thousand Jewish settlers (Central Bureau of Statistics 2015) in a settlement that has the status of an Israeli municipality and development town. It is situated adjacent to Hebron, a large Palestinian city and key economic hub with over 215,000 residents (Palestinian Central Bureau of Statistics 2016). An offshoot of Kiryat Arba, the remade Jewish Quarter was subsequently established in 1979 and lies directly within the historic Old City of Palestinian Hebron. It is a heavily militarized enclave of seven hundred residents including many children. The urban location of this offshoot makes its settler presence particularly volatile, resembling only parts of the Old City in Jerusalem. The paramount religious site in Hebron is known (variously according to each community) as al-Ḥaram al-Ibrahimi (Abraham's Sanctuary) in Arabic, Meʿarat ha-Makhpelah (Multiple Cave) in Hebrew, or the Tomb of the Patriarchs (a name given to the site during the British Mandate). Judeo-Christian and Islamic traditions all designate the area as the burial site of three key patriarchal couples, Abraham and Sarah, Isaac and Rebecca, and

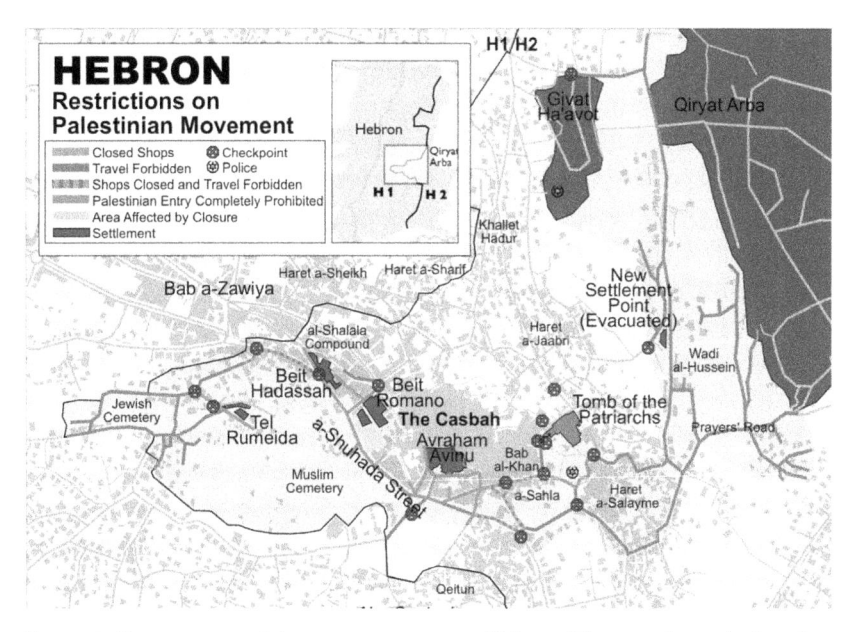

Figure 2. Restrictions on Palestinian movement in Hebron. The map shows the Jewish settlement of Kiryat Arba (Qiryat Arba, upper right), the Tomb of the Patriarchs (lower right), and the Jewish Quarter in the Old City of Hebron (Beit Hadassah, Beit Romano, Avraham Avinu, near the Palestinian market [the "Casbah"] as well as the reopened Jewish cemetery), and areas where Palestinian shops have been closed and travel forbidden. H1 is administered by the Palestinian Authority, and H2 is administered by the Israeli military. Used by permission of B'Tselem—the Israeli Information Center for Human Rights in the Occupied Territories.

Jacob and Leah. The Jewish settler presence in the Tomb of the Patriarchs has turned this site of convergence into a touchstone of violence. While the site functioned as a mosque from the seventh century until the Israeli occupation of the West Bank in 1967, it was subsequently partitioned for use as a synagogue and mosque.² Jewish settlers have also claimed and settled formerly Jewish residential areas near the mosque that were evacuated in the wake of 1929 anticolonial riots. Other provisional settler outposts and semipermanent housing have subsequently been scattered throughout the hilltop areas to the south known as the Hebron Hills.

After the June 1967 war, the Israeli government initiated settlement efforts to control the territories it now ruled. In Gaza and the West Bank, heavily populated Palestinian areas were placed under direct military rule, and border areas were settled for what the government deemed to be security reasons. Religious right activists took these official settlement efforts as an opportunity to realize their own theological ambitions in Hebron. Kiryat Arba (*Kiryat ʿArbaʿ*), the "fourth village" in the Bible, and its radical offshoot, Hebron's remade Jewish Quarter (*ha-Rovaʿ ha-Yehudi*), were both established illegally and then retroactively recognized by the Israeli government due to their religious value. In the wake of the 1993 Oslo agreements and 1997 Hebron Protocol, this H2-designated area has become an exceptional zone. Approximately 35,000 Palestinian residents still live directly under the authority of the Israeli military as they did during the pre-Oslo period, cut off from most of Palestinian Hebron placed under the control of the Palestinian Authority, without municipal services or adequate security protections.

Religion as an Ideological Formation

This ethnography approaches religion in settlement as an ideological medium rather than as a symbolic system in order to focus on its transformative and power-laden potentialities. My aim is to document the lived rather than merely textual aspects of Judaism in this particular context in order to highlight how its transformations legitimate processes of territorial expansion. This "ideological" designation also references an internal distinction that Israelis themselves make, distinguishing settlers who have moved over the Green Line (the 1949 Armistice lines) for religious reasons from those who have moved due to economic or quality-of-life incentives and in search of affordable housing.³ By ideology, I refer to the "amalgam of ideas, strategies, tactics, and practical symbols" used for realizing social and political change (Friedrich 1989:301). I distinguish this from the classic (Marxist) formulation of "ideology" as that which distorts an actual underlying truth, or its related Gramscian version as the way in which a ruling class not only justifies dominance but manages to win the active consent of those over whom it rules (Kertzer 1979:324). Rather, I give the term a different inflection, referring to a particular version of religious observance among those who advocate social change while relying on "asymmetrical or exploitative relations of power" to achieve their ambitions (Friedrich 1989:302). These unequal relations are fused with understandings of "difference" that tend to distort, obfuscate, and constrict possible "imaginings of the self" and "dialogic as well as other human relations" (ibid.).

The "ideological" here then implies thinking about the ways Jewish tradition has been particularized and funneled through the lens of settling. This narrowed or fundamentalist focus involves three further changes that are also useful for framing this study: the first is that religiously inscribed space, particularly the remaking of many Palestinian areas into a geography of biblical sites and origins, has been given a new significance in the construction of a distinct Jewish (settler) identity. Spatial reorganization has also resulted in a range of incremental practices included under the rubric of religion that link up with this process of inscription—including renaming, reenvisioning, and rebuilding. These practices in turn support and magnify resolute place-based attachments. The second shift is that these remade biblical sites, specifically in Hebron and within the Tomb of the Patriarchs itself, are being given a new centrality in Jewish observance, one that largely cancels out the exilic orientation of Jewish tradition. They give rise to a form of Jewish observance focusing on exact origins and specific graves to the exclusion of a more characteristic yearning for the messianic future. Third, the final change entails writing out the many historical convergences between Judaism, Christianity, and Islam reflected in the traditions themselves so as to eliminate possibilities for accommodating difference, while using Jewish observance and forms of direct violence in order to erase the presence of an existing Palestinian population.

These particular shifts in Jewish observance are understood to be ideological in the sense that they mask existing relations of power, having affective and embodied elements that play a role in shaping overtly political but also intimate decisions. They include modes of religious (settler) life that may not always have instrumental aims but that nevertheless have significant political effects (cf. Friedrich 1989). To reiterate, then, I distinguish my approach from that focused on either a symbolic or a textual analysis of tradition alone as well as that exclusively based on comparative forms of settler colonialism because of this synthetic religious and spatial character, and its relation to the context of a contemporary military occupation. If we are to take the ideological aspect of settling seriously then, as I intend here, we need to approach it not as a product of devout ideas alone, whether as canonical texts, on the one hand, or as religious ideas only furthering extractive colonialism, on the other. Instead, the ethnography shows that ideological settlement entails distinct practices, values, and communities that are oriented toward remaking much of a known Jewish ethical terrain and form of devotion while also appropriating Palestinian land on religious grounds.

Continuities and Disjunctions in Settlement

Labor Zionist and later Jewish refugee immigration to Palestine, the 1948 War of Independence/Nakba and Palestinian expulsion, the state's pronatalist policies, and military rule over remaining Palestinian areas played substantial roles in creating a Jewish majority in the Israeli state. These past iterations of settlement have been linked to processes of demographic change, both in establishing the Israeli state and later as a means of distributing a distinct national population throughout its territory. Settlement has also served as an Israeli security strategy, using populations to guard areas bordering on Arab states. Yet settling out of religious right devotion, as opposed to Jewish (ethnic) affiliation, was a later addition to the settler equation given that most observant Jewish communities initially defined themselves against a Zionist ethos. Observant Jewish communities motivated to settle for mainly theological reasons in what they deemed to be the biblical Land of Israel appeared on the political horizon only in the wake of the 1967 war. These ideological settlers began to espouse their distinct view of Judaism, linking it with nationalist territorial expansion during a period of popular exuberance after the 1967 war, when Israel tripled its land mass and took control of Jordanian-ruled East Jerusalem and the West Bank, as well as areas formerly belonging to Syria and Egypt (Sprinzak 1991, 1999; Segev 2008).

In its most basic sense, settlement as discussed here means creating an ethnic enclave within another population for the purpose of marginalizing its power and/or controlling its resources. This book focuses on the post-1967

Jewish ideological settlements that have been established for religious reasons within Palestinian population areas beyond the Green Line, Israel's de facto border. I concentrate on the settlements built adjacent to and within the Palestinian city of Hebron because this is where some of the earliest and most deeply ideological areas of Jewish settler residence were built. The term "ideological" distinguishes these settlements from those that are deemed to be overtly security-related and state-initiated built along the Jordanian border, or "economic" in that they attract residents because of cheaper housing.[4] While the divides between these different kinds of settlers have often been blurred in practice, I focus on those initially established for religious or ideological reasons alone in militarily occupied Palestinian Hebron because they were deemed to be exceptions to the security strategy pursued by the Israeli government at the time.[5]

Comparatively speaking, the term "settlement" comes from a broader settler colonial formation, where, as the scholar Patrick Wolfe (1999:209) aptly suggested, invasion is a "structure rather than an event" or a process that takes shape over the *longue durée*. This process entails resource extraction by establishing power asymmetries based on conceptions of social difference. Settler colonialism, in other words, features forms of "destruction that seek to replace" (ibid.), displacing populations from their territory. As a technique of both land acquisition and rule, it has an extensive history, appearing in the medieval writings of Machiavelli, the conquest of the Americas, westward expansion in the United States, Australian and Japanese colonialism, as well as many other contexts of resource extraction (Wolfe 1999; Elkins and Pedersen 2005). Understanding the influence of a distinctly ideological form of settlement in the Israel-Palestine case, however, requires thinking through the elaboration of a religious ideology as lived and practiced in the context of an occupied military zone, on the one hand, and thinking through its relation to a changing Israeli national project, on the other. It has a significant colonial dimension but includes other important drivers and determinations as well (cf. Ram 1999, Shafir 1999).

I therefore take the position that it is necessary to account for ideological settlement's specificity over the 1967 Green Line if only to trace its disproportionate social and political influence within Israel itself. What distinguishes all settlement in this context from its earlier periods is, first, that it is conjoined to an existing state that has never entirely defined its borders. The occupation of the West Bank on Israel's periphery created, in other words, an in-between zone, featuring a densely populated Palestinian area governed by a semi-permanent military administration. In this territory under occupation, multiple legal codes have been at play, private Palestinian property has been confiscated by decree, and its residents have been excluded from the rights of citizenship (cf. Hanafi 2009). Settlement added to this legal gray zone a more permanent social presence than what was presumed to be merely a temporary and finite military deployment. It contravened international law and made the Israeli military occupation of the West Bank one of the longest standing occupations in modern history. However, religious settlers have forged a distinctly hybrid modality of power that draws on religious authority and practice as well as legal ambiguity to propel social change. Viewed from the ground up, the subjective sensibilities and values of religious settlers are profoundly place-based in the sense of being resolutely attached to sacred sites and, emanating from these, linked to larger parcels of land distributed throughout a fragmented landscape (cf. MacDonald 2003).

Religion and Space

In studying the transformative potential of religion as an ideological medium, I pay particular attention to the role that religiously inscribed space plays in remaking or dividing up Palestinian areas. Given the scale of these spatial processes on the ground, I focus on the intersection of "religion and locality" (Knott 2005, 2009). Highlighting religion's relationship to the local, I analyze how settlers reorient Judaism to their social, economic, and geographic conditions as well as to Israeli military rule in Palestinian Hebron. In doing so, I analyze religiously inscribed space as that which has dynamic and power-laden qualities (cf. Lefebvre 1991; Foucault 1977, 1980; de Certeau 1984; Massey 1993).[6] Space then is treated as an important ideological medium of change that can be actively used to shape forms of domination rather than as just a container of settler actions.

Highlighting the significance of this religiously inscribed space, however, doesn't preclude its relation to temporality or the past. Rather, the focus on space also implies the ways conceptions of time "flow through space" and the way temporal reconfigurations of the past and biblical places of origin enable new power relations to take hold (Knott 2009:156). As this ethnography shows, time as well as space becomes increasingly fragmented through its reiterative and cyclical rendition, furthering settler practices of inscription. Settled spaces, then, tend to embody present orientations (the here and now) rather than future-oriented sensibilities and are thought to commemorate events that often repeat themselves—biblical, historical, the just past, and the ever present are temporalities that the settler imaginary stitches together to create a seamless sense of continuity and permanence.[7]

With respect to settling, this ethnography highlights how ideological settler uses of space herald a particular version of the past, which has significant social and political ramifications in the present. It shows not only how bounded spatial fields are made to invoke the Bible, but also how these cannot be disentangled from the religious investments and discourses shaping them, along with elements of social production and reproduction (cf. Knott 2009). So, for instance, the following events, which will be discussed in the ethnography, appear not as unique historical events in their own right but as reiterations lacking any historical specificity: the burial of the biblical matriarch Sarah, the burial of Hebron's Jewish victims of the 1929 massacre, the 1974 burial by Sarah Nachshon of her son Abraham (in the wake of sudden infant death), the commemorations and burials of post-Oslo victims of Palestinian violence, and an array of deaths from natural causes. All of these are lined up as a sequence destined to repeat itself. In this, little difference across time is recognized, producing a distinct sense of victimization and fatalism. Reiterative time also makes plausible the naturalized replacement of Hebron's former Jewish community, victims of the 1929 massacre, with that of a violent settler vanguard residing directly in Hebron, foregrounding ethnic similarities alone. It is worth noting that the focus on 1929 is a local settler memory that was all but erased in the national Israeli framework of memorialization because Hebron's former Jewish community was anti-Zionist, religious, and Arabic speaking (Feige 2001; cf. Cohen 2015).

Given the rubric of "religion and locality," this ethnographic approach is also distinguished by a preoccupation with small-scale devotional practices. By practice, I mean culturally patterned behavior that results in the spatial remaking of sites rather than preexisting social structures, rules, or cultural norms and that determines individual or collective action (Bourdieu 1977). This focus on spatial practice rather than on "materiality" or "environment" is intended to foreground contingencies, improvisations, tactics, and mishaps that figure into the remaking of space as an imagined biblical landscape, as well as adaptations of Jewish tradition itself (cf. Thift 2008; de Certeau 1984). My approach also stands in direct contrast to other classic frameworks that focus on the symbolic structure of religion (Geertz 1973) and its key concepts (e.g., the sacredness of land, messianic redemption) in the same way that

Figure 3. Archival photo of Hebron's Talmud Torah (traditional elementary school) taken in 1902 featuring (seated in the center) Rabbi Meir Shmuel Kastel (leader of Hebron's Arabic-speaking Sephardi community), Rabbi Rafael Franky, and Rabbi Hassan (chief rabbi) and Rabbi Tzarfati, who were killed in the 1929 riots, among other teachers and students. Central Zionist Archives, PHG/1023642.

analyses of language distinguish between an underlying grammatical system and pragmatic events of language use (Saussure 1983[1915]).

To argue from symbolic concept to action, as symbolic analyses often do, suggests that actors are essentially captive to their ideas. Their behavior, in other words, appears to be uniquely determined by ideas that withstand change. Rather than providing an interpretation of a religious-symbolic system that presumably determines settler behavior on the ground, then, I consider settler practices that shape a new geography as well as religious interpretations that often lead to violence. This focus on practice also allows me to consider the paradoxes and contingencies of settling—insofar as creating new settler strongholds or expanding old ones is often an incremental process that does not always follow a given plan. There is an important situational aspect to settlement that cannot be captured by textual and symbolic accounts of Judaism alone. Yet at the same time, one needs a broader sense of how ideological settlers actually invoke tradition to create their locations and destinies in order to see how radical religious commitments resonate with an evolving material reality.

Place Making and Devotion

Small-scale settler practices in Palestinian areas result not only in places deemed to have Jewish origins but also in resolute attachments. These deeply felt religious investments in places are often mobilized for political aims, supporting the transformative potential of religion as an ideological medium (cf. Cresswell 2004; Massey 1994; Tuan 1990; Casey 1993; Brauch et al. 2008). Moreover, sacred places in this context also have a strong ethnonational aspect. In both cases, the relationship between people and place is forged on the ground through practices seeking to transform a geography that is itself being reorganized more broadly by the state (cf. Weizman 2012). While salient ethnonational practices include walking, parading, intimidating, confiscating, vandalizing, destroying, and demonstrating through Palestinian areas in Hebron, they may also be combined with elements of Jewish textual tradition as a form of legitimation, revealing combinations that are both syncretic and malleable while nevertheless orthodox in self-conception.

Because deep attachments to specific sites are a central feature of a settler identity, I map out their (paradoxical) character.[8] Ideological settlement has become a way of lending a distinct order to the memory entailed in Jewish tradition, curtailing its dynamism, while narrowing possibilities for interpretation or revalorization. It shares features with Pierre Nora's *lieux de mémoire* in the sense that it actually erases more variegated forms of local memory in favor of preserving narratives of a homogenized ethnonational past deemed suitable to this distinct form of Jewish observance (Nora 1989). In addition to erasing this type of variegated memory, ideological settlement reconstitutes the diasporic and transportable features of Jewish tradition replacing it with precise biblical sites. It does so in a way that is more literal and place-based than other known versions of Judaism, or labor Zionism (cf. Feige 2001).[9] The literalism, then, of the religious settler project lies in the attempt to make places of *habitation* as sacred as the religious requirement of observing laws around marriage, diet, burial, and other commandments governing social life (*mitsvot*). In this manner, the legalistic aspect of Jewish tradition (halakha) becomes place dependent and enmeshed in actually inhabiting a material sacred geography (cf. Smith 1992).[10]

By focusing on place-based ties to religious sites and biblical regions (those that settlers refer to as "Judea" and "Samaria"), I distinguish resolute settler attachments from either the territoriality espoused by a sovereign Israeli state or theological concepts that view the Land of Israel as more of an aspirational terrain than a thing to be directly possessed. In short, resolute settler attachments remain small-scale and bounded, and though local loyalties and communities often coexist with those of a unified national culture (Appadurai 1996; Lomnitz 2001), these in particular cannot be encapsulated in a standard nationalist account. Settler investments in places like Hebron, in other words, continue to foreground the religious locale as the predominant basis for building the kinds of solidarities that often exclude even Zionist forms of belonging. They are shaped in relation to a tiered and asymmetrical social field within an occupied area and are therefore ultimately more focused on direct colonial or interpersonal relationships in self-enclosed and bounded insular worlds rather than on participating in a wider imagined (national) community.

Tradition and Change

This ethnographic emphasis on religious place and practice is an important way of considering changes to and distinct interpretations of Jewish tradition that are produced in relation to the many spatial practices that underwrite settling. It raises the question of how we can critically engage with these transformations and their implications for religious right social change. As a way of responding to this question, I draw on Talal Asad's approaches to the social and power-laden dimensions of religion (based on his scholarship focusing on elements of Islam and Christianity), while at the same time resurrecting a number of critiques that are disallowed by them. That is, I aim to unpack some of the exploitative aspects of devout settler practices ushered in under the guise of Jewish authenticity while nevertheless recognizing the excesses of secular critiques, which tend to characterize religious orthodoxies as "irrational," "backward," and "threatening" by their very nature.

In order to explain my approach further, let me begin by mapping out key convergences and departures from those of Asad on religion. He understands religious traditions as a way of situating the self with respect to the past from the vantage of the present and contrasts this with secular modern tendencies to look primarily to the future as a form of identification and orientation (Asad 1993, 2003). Asad also reminds us that using the richness of tradition in this way is not the same as being backward. These insights are valuable because they show how uses of tradition entail change; traditions in his view are not "invented" (Hobsbawm and Ranger 1983) but always transforming, since they must respond to contemporary challenges in the present, both internal and external, while trying to maintain a general coherence among key defining concepts (Asad 1993). Grappling with changing applications and understandings of a religious tradition for Asad also means rejecting scholars' use of the category of "fundamentalism" (cf. Marty and Applebee 1991), which he sees as critical of piety more generally, based in part on the accusation of either politicizing or introducing false changes to religion. So Asad would reject the accusation that radical Islam is not true Islam or that settler Judaism is not true Judaism. Using the term "fundamentalism" as a way of characterizing these changes, he and others argue, misrepresents the changing nature of all religious traditions as well as their power dynamics and highlights the lack of reflexivity found in secularist critiques that project irrationality and violence onto religious views alone.

Moreover, Asad rightly points out that there is an important social dimension to religion that cannot be reduced to beliefs alone. He has incisively alerted us to the fact that power and hierarchy shape all religious communities. Asad (1993) also highlights the conditions under which religious "truths" are held to be valid, emphasizing the power dimensions of pedagogy, disciplining the body, and submission to religious authority. Yet in defending a place for religious perspectives and modes of resistance, Asad tends to underplay consolidations of power being ushered in under the sign of religion and the way religious authority can be impervious to critical engagements apart from those offered from within a religious community. Also, in terms of Asad's explorations of power, he underplays the significance of space (emphasizing the body instead) as the primary medium through which religious subjectivities are shaped and relations of power expressed.[11] The social hierarchy that is congealed in the spatial ordering of the built environment, for instance, significantly shapes the ethos of sacred sites and subjective attachments to it in a settler context. Neither involves submission or compliance alone, but they entail direct (and antagonistic) engagements with difference that require the submission of others.

This ethnography's emphasis on spatial practice, then, builds on many of Asad's critical insights but ultimately takes them in a different direction. First, it looks at the manner in which an entire spatial field gets entangled in settler forms of religious knowledge and discourse (cf. Stump 2008). Second, it focuses on a range of heterodox practices (e.g., property takeovers, parading, trespassing, harassment) that some might object to as not religious even though they are often enmeshed with an evolving ideological formation. Such practices are often religious in the sense of invoking concepts within a tradition, yet mainly colonial in their impacts, underwriting land expropriations by individual or settler groups who often operate under the auspices of the state. Finally, it embarks on a critical engagement with "orthodoxy." I depart from Asad and others writing in this vein (Dalsheim 2011; Mahmood 2005; Brown 2013), then, by focusing on a range of exploitative practices, views, and sensibilities shielded by religion while nevertheless remaining sympathetic to attempts to decenter the primacy of secular rationality as normative.

Toward this critical engagement with settler orthodoxy, I take the spatial dimensions of religious practice to be a key domain of power. By examining the formation of an exclusive biblical geography on the ground and its pointed articulations with the state and its military, this ethnography attempts to offset the excesses of a (far-right) religious realm insulated from challenge by virtue of its material inscriptions and forms of naturalization. It also attempts to highlight the realm of less formally transmitted religious sensibilities in order to analyze how a range of settler devotees, including new converts and those not all that well versed in Jewish tradition, take part in claims to origins and place that evolve and change. Religious modes of (far-right) power that marginalize already precarious others, refuse dissent, and thrive on inequalities attempt to replace existing hierarchies with more authoritative forms of exploitation. Building on these concerns, this ethnography also evaluates the changes to Jewish tradition and practice introduced by settling—namely, the narrowing of an interpretive tradition and the colonial implications of a version of Judaism that serves to prolong a long-standing occupation.

Much of the scholarship on Jewish fundamentalism, in contrast, maps out the broad contours of change in a way that is mainly textual in its emphasis, exploring Judaism's traditional messianic focus and its encounter with the historical founding of the Israeli state (Ravitzky 1996; Attias and Benbassa 2003). Its overriding concern is with how Judaism accommodated the modern Israeli state—de-emphasizing a quietist anticipation of the future (e.g., waiting for the Messiah) while replacing it with an emphasis on human agency. Moreover, the scholarship emphasizes a renewed focus on human action as the catalyst for "redemption" popularized by the religious modernizer Rav Avraham Itzhak HaCohen Kook (1865–1935). Redemption (repair from exile), or *ge'ulah*, according to Rav Kook the elder, was no longer to be a miraculous event solely directed from above but would be realized through human action and signaled through the signs of divinity expressed in the material world (Ish-Shalom 1993:233). This position in turn prefigured the views of his more hawkish son, Zvi Yehuda Kook (1891–1982). Both were taken to be the spiritual forefathers of the Gush Emunim, the religious settler movement of the midseventies, which set up many of the first ideological settlements in the West Bank (Shahak and Mezvinsky 2004; Zertal and Eldar 2009). From A. I. Kook's writings, arguing that the process of achieving a divine end was as imbued with "divinity" as the end itself, came his son's assertion, some sixty years later, that all of Israel's territory was sacred, and no part of it could ever be forfeited in a peace agreement (Kook 1991). Messianic redemption was therefore reenvisioned as settling in the here and now and entailed taking steps toward rebuilding the Third Temple and returning the secular Jewish masses to observance.

While this literature captures important transformations in Jewish tradition, it overlooks changes in values, sensibilities, modes of understanding, and emphases that have been introduced through practices of settling and its relation to violence. This ethnography takes shifting attitudes toward *messianism* into account but focuses more directly on the ways settlers invoke and use Jewish tradition in routine practices for a project of social restructuring.[12] As an analyst, I approach "religion" as a lived and historically specific tradition rather than as a transhistorical or universal category and defend the potentialities inherent in religious lives (Asad 2003) while looking at the construction of a distinct settler iteration of "authenticity." In terms of Jewish observance, then, this ethnography refers to practices and modes of understanding, both oral and written, shaped in relation to canonical Jewish texts and Jewish law, including the Bible, or Tanakh (Torah and the remaining books of Nevi'im, and Ketuvim), as well as to classical rabbinic interpretations of halakhah (religious law), passed down orally and codified as the Talmud (Mishnah and Gemara). In addition to scripture and law, by "religious" I also refer to the heterodox sensibilities and practices that shape the character of devout settler place attachments as it has evolved in Kiryat Arba and Hebron. In its lived form, settlers circumvent Judaism's admonition not to conquer and blur the core divide in Judaism's legal tradition between hypothetical laws that are only to be implemented in the future and those intended to be observed in the present. This shift is arguably more significant than that related to messianism alone.

State, Nation, and Settler Relationships

In addition to the dimensions of religious ideology discussed above (spatial, place-based, tradition, and power), this ethnography also examines the role of the state because of the critical role it plays in creating the infrastructure for settlement and the dynamic tension between state-initiated programs and popular settler aims on the ground. Without the state overseeing the development of a large infrastructure and allocating financial resources to it, including roads, utilities, subsidies, and security measures, settlements would cease to exist. Yet it is also the case that religious settlers have their own agency and intentions that are not entirely reducible to state initiatives and programs. Exploring settlement from a state-centric perspective alone precludes understanding how state directives are being taken up or ignored on the ground by average settlers. Moreover, ideological settlers have tried to develop their own sphere of influence, pushing the boundaries of state and military authority. While the state matters, in other words, this ethnography reveals that it enables but does not directly oversee many of the routine practices of ideological settlers on the ground. In short, we are left to grapple with the paradox of a state-sponsored ideological formation that seeks in the long run to empty the state of its authority and institute theocratic rule in its place.

One prevalent view is that the local set of attachments established via settlement is a direct continuation of the Israeli nationalist project and that through settling one merely extends state control as well as expands the boundaries of the nation. In short, from a macro perspective, it is easier to claim that all settlers see their particular locale as a vehicle for participation in a wider set of national affairs. Their insistence on establishing biblical memory sites might even be viewed as a way of apprehending a more abstract political project from the vantage of direct experience: the locality is in effect a bridge to a larger polity, where being at home in a settlement translates into larger loyalties toward the nation. This is the *Heimat* model of national belonging, which emphasizes the local but is not averse to participation in a wider political community (Applegate 1990:13). Instead it represents a contemporary form of imagining through the vantage of one's own deeply rooted historical reality (ibid., 3).

By contrast, I take the position that the locale in ideological settlement has evolved as a platform for breaking away from the national project. As a settler once quipped, "Do not insult me by calling me Israeli." He intended to convey the idea that the nation's presumed Jewish character had been diluted beyond recognition and that he was engaged in a revitalization of it. Settler sites of religious origin and biblical locales admitted into the Israeli national arena often compete with and overtake other sites of memory that have been central to the heritage of the nation. In this respect, Hebron stands in tension with the more pluralistic character of Jerusalem. It is about championing an alternate form of authority, stemming from a set of religious obligations rather than rights, a prioritizing not of territory but of critical biblical sites as anchors of memory. And this involves a form of political affiliation that moves far more easily between the local and the transnational as a way of appropriating the national than extending national rule alone. In short, I am skeptical that ideological settlement is simply a case of putting nationalism on more solid ground by adding a fervent religious layer to its comparatively secular expressions.

Fieldwork in Kiryat Arba

The idea for this ethnographic account first took shape during the 1993 period of the Oslo talks, a brief interval of optimism. At the time of the Oslo accords, which were significant for being the first formal agreements made between Israel and the Palestinian Liberation Organization (PLO), many considered themselves to be witnessing a historic turning point. It was the first time Israel and the PLO recognized one another and the first time Palestinian rule was to be granted over areas that had previously been ruled directly by the Israeli military. Oslo's flaws, however, appeared early on. Accounts of the period point to the lack of vital decisions made over Palestinian refugees, borders, statehood, settlements, and Jerusalem as the primary reason for the failure (Beinin and Stein 2006; Golan 2013; Khalidi 2006). Others fault the negotiation process itself, which was carried out in back channels while international attention remained fixed on multilateral initiatives in Washington (Savir 2010; Khalidi 2013). Still others lay the blame at the hands of individual leaders, both Israeli and Palestinian, and their personalities (Falk 2013). On the Palestinian side, Oslo's collapse led to the rapid deterioration of an already precarious economy, greater Israeli military interference in Palestinian lives, and the creation of a Palestinian Authority many considered corrupt. On the Israeli side, as the security situation for civilians worsened due to suicide bombings, right-wing hard-liners gained prominence. Promising the Israeli public greater security, the Likud ran on platforms that emphasized aggressive military action together with the building of an extensive security barrier. Oslo's failure also precipitated an accelerated Israeli landgrab using the vehicle of settlement expansion (cf. Mansour 2001).

I began investigating ideological settlement in 1994, in the period just after the Oslo Accords. The period was marked by vitriolic settler protests, on the one hand, as well as several Palestinian suicide bombings within Israeli cities, on the other. The assassination of the Israeli prime minister Yitzhak Rabin occurred in the wake of these many months of unrest. Just one day after the assassination, the owner of a studio in Kiryat Arba phoned me in Jerusalem to finalize my lease. This struck me as odd because the entire country was in a state of shock and mourning, and not once did she refer to the assassination. She proceeded, in an entirely businesslike fashion, to sort out the details of my move, emphasizing the benefit of living in a "desirable" location—the clean air, lack of congestion, lovely views, tight-knit community, and affordability. In retrospect, I imagine she was worried about finding a tenant because the tide of Israeli public opinion had turned against Kiryat Arba. Yigal Amir, Rabin's assassin, maintained ties with settlers in Kiryat Arba, and in that period it was directly linked with his religious extremism.

I got off the armored and steel-grated bus that traveled from Jerusalem directly through the Palestinian town of Beit Jala, past the refugee camp of Deheishe into the adjacent town of Halhoul, until it stopped at the center of Kiryat Arba. It was the middle of the day and I arrived with two large overstuffed bags in order to carry out a research plan that I had conceived of in Chicago. The actual settlement felt pretty much dead and deserted except for a few kids playing in rusting playgrounds, their empty swings and ladders straining between the concrete and stone blocks of older prefab buildings. As I approached the four-story complex I would live in for much of the year, a three-year-old rode up on his faded plastic bicycle, seat low to the ground. I noticed, in spite of his refusal to say much, that he followed a few paces behind me, making a great effort to pedal fast enough to keep up. When I turned toward the entrance to the apartment building where his family lived, he broke his silence and pointedly said, "You're secular," referring to me as *ḥilonit*. It is true that I was wearing jeans rather than the long flowing skirts he was used to seeing, but it took me by surprise that even this young boy with a knitted skullcap and fringe slotted me so quickly. He carried on: "And your mother, father, sisters, and brothers are all secular too!" "Right," I said with a certain didactic enthusiasm. Had I known more, I might not have been so quick to agree. Calling someone "secular" has a deeply pejorative sense, and as I got to know a few people in the settlement, they preferred that I call myself *masorti* (traditionally observant). *Masorti* meant, in their view, that I was not strictly observant but open to a religious point of view—that I would light candles on the Sabbath, keep kosher, and cease to work on the Jewish holidays, even if less rigorously than those who were devout.

I walked through the door of my rented studio, which had previously been inhabited by an Israeli soldier fulfilling his required service. It was a stark space with one window and a small cooking area. I imagined I would gain a better grasp of how settler ideology operated, though in truth this aim seemed rather abstract once I had arrived. More tangible was the physical layout of Kiryat Arba. It had a central core of four-story buildings, surrounded by Palestinian agricultural fields, which were farmed infrequently as it turns out, because their owners were prevented from accessing them. Then in view of these agricultural areas, there were other Jewish residential neighborhoods—newer parts of the settlement perched on hilltops overlooking the heights, including Givat Harsina, built and named after the

Figure 4. Newer extension of Kiryat Arba encroaching on a Palestinian agricultural area, 2007. Photo by author.

pullout of the Israeli army from the Sinai Peninsula in the 1980s and Eshmoret Yitzhak, a newer extension.

The settlement was a patchwork of enclaves separated by green spaces. Cyclone fences, topped with large coils of barbed wire, enclosed the settled places. I was struck by the circularity of the roads in an older section of the settlement—the main road went around the settlement circumference, and the other roads seemed to circle into the main road, giving one the illusion of going somewhere, but then not. One ended up at the point where one began. These internal loops eventually led to a choice of one of three exits: the main gate off the road (Route 60) to Jerusalem; the western one by which the hourly bus and cars traveled on an unpaved road through Palestinian residential areas to reach the Tomb of the Patriarchs; and a little used gate near an industrial zone, apparently accessed more frequently in the past by Palestinian laborers employed in the factories within the settlement. Since all settlements are cut off from the pulse of life that surrounds them, they mainly tend to have a static quality compared to either cities in Israel proper or the Palestinian towns and villages that immediately surround them. In the middle of the day, with many of Kiryat Arba's residents away at work, the settlement felt like a ghost town. In the evenings and on weekends, people gathered together in public areas and the stark housing came to life.

The bus, which arrived hourly from Jerusalem, turned out to be an important lifeline. Walking between neighborhoods wasn't done because it required traversing Palestinian areas on foot. Most settlers either waited for the bus to get a two-minute ride to other settled sections or informally waved down a passing car. I once walked between these areas marking myself as a distinct outsider. A founding member of the settlement whom I had interviewed the day before saw me on the road and instantly stopped to give me a lift. I asked him about the patchwork character of the space, and his response was "Aren't the vineyards nice?" My errors and inquiries were closed off by pleasantries. "So green," he remarked. I had the sense of being in a small town where people took note of even slight transgressions. This attention to detail was coupled with a decided lack of seeing any Palestinians who lived across the fence. "I see them, but hardly see them," one Kiryat Arba settler once remarked in passing. But residents of the settlement did keep an eye on me.

Some settlers I spoke to remarked that they were relieved I was not simply another journalist passing through. Yet the small-town aspect of the settlement and the troubled times created an atmosphere of suspicion and paranoia. Why did settlers speak with me? Perhaps it was the palpable monotony of daily life, the intent to convince an outsider of a religious point of view, or the hope of bringing a secular analyst into the fold. A synthesis of these experiences, together with my observations and formal interviews form the basis of what I present here as the dynamics of an ideological settler formation unfolding—one of seeing selectively, of using possibilities afforded by religious devotion to create resolute attachments, of legal gray zones, complicity with violence, and the search to recruit distant Jews. These aspects of the everyday deserve our attention because ideologically, emotionally, and even culturally an ideological settler view of the world is marked by its circumlocutions, erasures, as well as devotions, particularly in light of its detrimental impacts on the presence of an adjacent Palestinian population.

Chapter Sequence

The research has been conducted in short and long stints for a long period. I was neither well received nor shunned as a researcher, although I am grateful to those who sought to instruct me in religious and political matters in spite of our resounding differences. My informants believed that I had been brought to this study for a different purpose from the research interests I thought had motivated me, namely, that of fulfilling a Jewish destiny. Others took a more pragmatic approach and felt that it was worth making their point of view known to a researcher, particularly because of the hostilities they faced in the international context. I made few friends, and the friendships that did form were partial and guarded. Settlers often saw themselves as under siege and targets of government surveillance, and some found it easy to classify my research as part of wider aims to monitor their actions. Others, however, were taken by the novelty of a visiting anthropologist and, while hosting guests and family on holidays, would announce in earshot that an anthropologist was present.

The chapters are arranged thematically and analytically with attention to the chronology of events as I encountered or recorded them. In sequencing the chapters as I do, a key concern has been to provide an understanding of the ways religious ideology gives rise to violence and to outline the legal and security background for it. I begin with spatial practices and later examine the Goldstein massacre in order to disrupt the immediate association between religion and violence that prevails in discussions of "fanaticism." I also highlight adaptations within Judaism that take place, invocations of tradition that align with settler aims, as well as the entanglements of spatial and religious elements in daily settler life.

The first chapter, "Orientations," is composed of three perspectives on religious settlement that evoke the different populations living in Hebron (ideological settlers, soldiers, and Palestinian farmers), showing the ways their views coexist as separate but intersecting realities. In a settler tour, a guide points to the presence of Jewish origins in Palestinian areas by invoking the Bible while redirecting the gaze, renaming places, masking human toil, and inventing genealogies. This contrasts with the views of a religious Israeli soldier turned conscientious objector who has also served in the area and those of a Palestinian farmer living adjacent to the settlement. The soldier focuses on the way a religious settlers' actions place the lives of fellow soldiers in danger and how the occupation skews a soldier's "ethical" choices. A Palestinian farmer discusses how he and his family have been impacted by ongoing acts of settler violence and the ways this violence occurs in tandem with the military's destruction of property, revealing a double subjugation. Though not all-encompassing, these points of view are intended to sketch out telling perspectives that one might readily encounter in Hebron's complex social field.

Chapter 2, "Between Legality and Illegality," provides a brief historical overview of the claims and practices that formerly paved the way for ideological settlement to take hold. Emphasizing existing relationships between settlers, the military administration in Hebron, and officials in the Knesset, the chapter explores how the first ideological settlers to enter Hebron capitalized on what they deemed to be Jewish "origins" within the legal gray zone of the military occupation. It shows how they established their presence by getting military permission to celebrate the Jewish holiday of Passover and refusing to leave. In this case, Passover initially appears to be the common ground uniting settlers, soldiers, and elected officials. Once established in the area, however, settlers begin to use Jewish observance as a means of bypassing military and state authority, making their temporary status in Hebron permanent. Not only is the ambiguous legality of the occupation, its "state of exception," relevant here but so too are the political divisions that lead to accommodation to a settler presence, in addition to the virtual lawlessness of the occupation that allows settlers to operate with impunity. Their invocation of Jewish origins and religious commitment seems to solidify ties with the state, but the difference between compliance with the law and undoing it is often very narrow. Settlers are therefore able to reinscribe territorial boundaries, push legal limits, and enshrine religious values in ways that establish their own distinct arena of control, albeit on religious grounds.

The process of actively shaping space and negotiating new boundaries is further elaborated in Chapter 3, "Motherhood and Property Takeover." The chapter uses the lens of gender, specifically cases of protest featuring religious settler women and children in the mid-1970s, to analyze how maternalism and motherhood were deployed for political ends and continue to be a key strategy. It features the ways women claim property in Hebron on ethical and historic grounds as they move out of their homes, recreate domestic settings in Palestinian areas, and use bonds with children to establish ties to designated Jewish areas and biblical sites. As in other cases of fundamentalism, the gendering of everyday life is central to this new ideological formation insofar as it is intended to counter the changing roles of women in comparatively secular Jewish contexts. The spectacle of women as mothers protesting demonstrates one of the ways ideology gets embodied and internalized. In protest, as in the home, cycles of pregnancy and childbirth often produce the ties that bind, while serving as the basis on which domesticated spaces using a matrilineal logic are being deployed to take over Palestinian property.

Chapter 4, "Spaces of the Everyday," focuses more directly on the inscription of an exclusive ethnic logic in spaces and practices that are being shaped by a devout settler presence in Hebron. It analyzes solidarities as they are worked out in a series of antagonistic social encounters between settlers and Palestinian "others." The chapter then points to ways in which settler bonds and ethnic sensibilities are further enabled by distinct (re)readings of Judaism's legal tradition and practice. The ethnography focuses on incidents where settlers storm through Palestinian space, engage in daily transactions that are marked by nonrecognition of the other, and participate in threatening exchanges over the fence that secures Kiryat Arba, separating it from Palestinian agricultural areas outside the city of Hebron. It also highlights the ironies, vulnerabilities, and impossibilities of creating a virtual world of exclusivity. Focusing on how spatially segregated areas and congealed notions of difference operate, the chapter then moves to a discussion of settler violence.

Having considered the micro-level clashes and disputes that multiply around boundaries, Chapter 5, "Religious Violence," explores violence in the context of the partitioned space of the Tomb of the Patriarchs. It analyzes the 1994 Goldstein massacre, a key incident of settler violence that takes place inside the Tomb, by drawing on government inquiries, interviews and observation. It details the actions that went into the partitioning of the mosque and the ways in which violence inheres in the incremental remaking of sacred space. The chapter also shows how Jewish belief and practices come into conflict with Islamic worship by focusing on graves and saint worship to the exclusion of other dimensions of Judaism. While many scholars point to the signing of the Olso accords as the event that triggered the Goldstein massacre, this chapter offers a more process-based and spatial account of violence. It features conditions of the everyday, showing that routine acts of settler harassment give rise to an ongoing relationship to violence and shape ideological views that favor the use of force. Not only are settlers armed, but they often work with soldiers or serve in the military themselves. As spaces of religious settler control proliferate, and Palestinian residential areas are caught between overlapping arenas of settler and military rule, so too do opportunities for settlers to transgress boundaries increase, blurring the civilian-military divide while appropriating military violence to increase religious domains of control.

The sixth and final chapter, "Lost Tribes and the Quest for Origins," transitions away from sacred place to the ideological settler search for Jewish origins in an ever-expanding diaspora with the aim of bringing "original" Jews back to their presumed place of origin in Hebron. The chapter highlights the diversity of Kiryat Arba's population—while featuring settler attitudes toward Ethiopians, Russians, and the Bnei Menashe, as well as many converts—exploring the ways settlers bypass the law and use rabbinical authority as a cover for it. In this process, presumed "native" forms of Judaism are found in lost tribes that mirror the contemporary settler self. The irony is that the many cultural and linguistic differences introduced in the process of immigrant recruitment turn out to be as great if not greater than the cultural differences that exist between ideological settlers and their Palestinian adversaries. Cultural and social differences turn out to be less significant than matters of loyalty and ideological affinity. The chapter focuses on alienations that come out of immigrant recruitment and integration into an ideological settler community and the ways that internal differences are actively negotiated in a process religiously referred to as "exilic ingathering." The push for conversion to Judaism, inimical to the known tradition, is ironically championed as a means of integrating many nontraditional Jews from remote places into an evolving settler formation.

The conclusion, "Unsettling Settlers," begins with the social dislocations that came about with the 2005 withdrawal of settlers from the Gaza Strip and moves to the subsequent reentrenchment of the second-generation of ideological settlers, the so-called "hilltop youth," in the hills of the Hebron area. It details the growing anarchistic and individualistic trends in religious settlement and the mainstreaming of religious settler values once seen as marginal. In doing so, it attempts to grapple with the legacies of a religious right rise to power in the Israeli political arena and beyond.

In sum, this ethnography aims to illustrate the changes that distinguish and define ideological settlement in terms of practice and sensibility from other settler iterations. It shows religious settlement to be a heterodox synthesis of religious practice and politics in a military zone enabled by existing structures of exploitation and the occupation as reflected in each of these chapters. Settlers champion permanence and habitation, holding onto seized locations as a way of forging a distinct kind of peoplehood. Under the mantle of religious authenticity, land can no longer be apprehended in its symbolic or mystical form, having redemptive potential. Nor can it be seen as mainly an ethical terrain that unifies people through shared values. Rather, it is a thing to be possessed and, in this regard, has been reduced to profane property. In the following pages, then, we encounter a distinct way of settling and set of Jewish practices that sustain it. But to more fully understand the context in which they operate, we need to start with an investigation of how ideological settler views and practices compare and contrast with those of others residing in the area. To do that, we move now to the first chapter. It sets out three distinct West Bank experiences and the power dynamics that inform them, as well as the ways these either intersect with (in the settler-soldier interface) or push against (in the Palestinian case) the singularity of an ideological settler vision.

Chapter 1

Orientations

To be a devout Jewish settler is a deeply contradictory project—from an ideological vantage, it requires reorienting exilic Judaism and its broad textual tradition toward specific sites that are deemed sacred. Not only does settling religiously emphasize actual places over the biblical Land of Israel, but more narrowly, it entails drawing precise correspondences from textual passages to sites in Palestinian areas where biblical events are deemed to have occurred.[1] In these attempts to remake the fabric of Jewish social life and marginalize a Palestinian presence, ideological settlement necessarily does away with many of the diasporic elements of Judaism and remakes tradition itself. Whereas in its exilic mode, Judaism relied on texts that could be carried, times rather than places of observance, shared language, and a set of religious laws in order to create commonalities in the absence of territory, in this settler version, observant practices that potentially bind people to place become far more valued. A great deal of remaking, reorienting, and even rupture, then, has gone into reinterpreting Jewish tradition through the lens of settling and endowing it with a sense of religious obligation.

This chapter explores the key dimensions of change in Jewish thought and practice that have provided religious rationales for settling in Palestinian areas. It focuses on three concurrent kinds of change, making the case that these are in fact the most significant for understanding an authoritative ideological formation in the making. These dimensions include the ethical realm of moral behavior in Judaism, its relation to place-based forms of residence, and the treatment of other non-Jewish residents that follow from this spatial (and territorial) turn (cf. Fonrobert 2009). Jewish settlers, then, are not simply messianic believers, shaped by a pragmatic orientation toward the coming of the messianic era as many allege (Aran 1991; Ravitzky 1996; Inbari 2009;

Taub 2011); rather, as believers, they are engaged in far more extensive and complicated forms of remaking. This deeper reworking of Judaism and the social life that follows from it entails the wholesale reorientation of its primary texts, ethical obligations, and rabbinic interpretations, in effect narrowing and particularizing the tradition's interpretive possibilities. Biblical places, boundaries, and ownership of land are some of the obvious emphases that settlers bring to Jewish observance, and these take on a value that overrides concerns with peoplehood and future-oriented messianic longing. Moreover, ideological settlers have added to their place-based emphasis a concern with seeking out and finding hidden or erased origins. Origins as much as observing religious obligations and laws come to define and orient the "authentic" Jewish community. In practice, then, devout settlers have emphasized the significance of a growing number of sites of origin, as well as the routes to them, in order to expand into key Palestinian areas.

As a way of introducing some of the critical themes of the book and focusing on the ways Jewish tradition is being reworked, I begin by reconstructing a tour of Hebron led by a Jewish settler, showing the way it invokes and applies elements of Judaism to claim Palestinian land. These settler claims are then juxtaposed with the perspectives of a dissident Israeli soldier who served in Hebron and those of a Palestinian villager whose farmland lies near Kiryat Arba. While the three tours through the Hebron area actually took different routes and occurred at different times, I discuss them together in order to highlight the ways a settler's religious claims are often reinforced by the dynamics of the military occupation. As parts of a composite whole, these different tours and perspectives are intended to both examine and disrupt features of an ideological religious vision that is becoming increasingly hegemonic.

Reframing Scripture

To provide a better sense of what settling religiously entails, I begin by turning to a reading of the Bible (Torah) that Hebron settlers typically champion as authoritative. Again, I am interested in the way the Bible's diasporic elements are routinely read out or reframed, and I consider it important to ask: What features of the text are being emphasized so that settling Palestinian Hebron appears to be biblically mandated in the contemporary moment? While Hebron has long been considered sacred to Jews and while there is ample documentation that an Arab-speaking Jewish minority lived in the city during the Ottoman period, Jewish settler activists radically transformed land adjacent to and directly within the Palestinian city after their arrival in 1968. If Judaism is actually a multivalent and multilayered form of observance, requiring a variety of texts to be read in relation to one another in order to draw out its key values through comparison and contrast (Fishbane 1998, 2012), how have settlers made it more singular and used the Bible as a charter for their practices of expansion? How has Judaism's legalistic emphasis on "doing" been reoriented to enable land takeovers? Moreover, how have interpretive debates that have taken place among rabbis or sages from the Middle Ages to the present been transformed by settler applications of these in order to erase the presence of Palestinians?

Let us begin with the particular reading of a biblical story often cited by militant settlers as a justification for settling Hebron. The passage appears in Genesis 23 as the "Life of Sarah," a prototypical story of Sarah's death and Abraham's purchase of a burial site for her. Abraham, a stranger (*ger ve-toshav*; lit., "a resident alien") in Canaan, approaches Ephron, a prominent Hittite and landowner, in order to negotiate the purchase of a field containing a cave, the Cave of Makhpelah, so that he can bury his wife. Abraham says, "I am a resident alien among you; sell me a burial site, that I may remove my dead for burial." In the passage, Ephron the Hittite responds that he may take it without paying, but Abraham nevertheless insists on purchasing it. The transaction between them is conducted within the hearing distance of the entire Hittite community and gets cemented by their respective social obligations to it as well as to one another. Their verbal agreement, then, has a public dimension and this helps make the land deal between individuals of two different peoples binding: Ephron says to Abraham, "A piece of land worth four hundred shekels of silver—what is that between you and me [*beni u-venkha mah-hi*]? Go and bury your dead [*ve-et metkha kavur*]." Hebron's contemporary Jewish settlers allege that the burial cave purchased by Abraham for the full amount is evidence that the contemporary place, where the Ibrahimi Mosque (al-Ḥaram al-Ibrahimi) now exists, is their exclusive property to this day. They claim to be the original owners and that the Jewish burial caves lie far beneath the Muslim cenotaphs above ground. For religious settlers, the presence of the mosque does not point to a convergence of Judaism and Islam but rather the usurpation of their original sacred Jewish site. They claim a right to worship directly within the space of the Ibrahimi Mosque because they claim it is built over bequeathed Jewish property. By extension, they also claim a right to expand into other Palestinian areas of Hebron because of the historical Jewish community that was massacred in the 1929 riots and forced to leave. What is clearly absent in their project of "renewal" is not only a sense of the interrelatedness of three traditions—Judaism, Christianity, and Islam—but the often syncretic ways Judaism has been lived in relation to these other traditions in history.

Casting aside the issue of being able to precisely map the topography of the Bible onto the contemporary Palestinian landscape or whether Abraham is a historical figure (biblical scholars consider him to be mythical), the settler claim depends on the idea that Judaism has a more authentic right to the site than do other religious traditions that emerge later in relation to it. Theirs is not, in other words, a pluralistic stance that recognizes multiple truths or the veracity of other traditions that intersect and come together in the figure of Abraham. Rather, Jewish settlers take as a given the need to pry these convergences apart and order competing claims along a linear time line. Christianity and Islam are thought of as later and less authentic additions. Ideological settlers are therefore concerned with returning to and restoring a form of authenticity deemed lost—not just a settler preoccupation, to be sure, but one that has been applied to the specific cause of settling in complicated ways. They also believe that as a Jewish vanguard (to the exclusion of other versions of Jewish observance or other Jewish communities in Israel and beyond) they have a right to stand in as the direct "inheritors" of Hebron for all other Jews. In other words, settlers emphasize that this property has literally been passed down to them across the ages (they are its heirs as the reflexive Hebrew term for "inherit," *hitnaḥel*, suggests) and that all Jews must eventually return to a place bequeathed to Abraham's descendants. They therefore claim to hold onto and inhabit property in Hebron not only for Jews in Israel and the direct descendants of those who actually lost property in the 1929 riots but for those living abroad in the diaspora. And yet other potential readings of Abraham's purchase highlight the settlers' interpretive reorientation of the Bible and show how it has transformed Judaism's values around property and relations with non-Jews.

Much of the scholarship on diaspora mentions burial as a common dilemma (Levy 2001; Levy and Weingrod 2004; Ho 2006). Individual members of a diaspora are inevitably forced to create geographic ties when burying their dead, giving rise to competing loyalties. In other words, burials spur attachments to a grave as a site of remembrance, and the grave competes with deterritorialized self-definitions that do not depend on any particular locale

(Levy 2001; Gonen 2004). Where, in other words, is it fitting to bury a person who has lived out his or her life within a diasporic context (cf. Clifford 1994; Huyssen 2003)? Anywhere and nowhere are equally sound answers. One could readily read Abraham's purchase of Sarah's burial site as working through this dilemma; a resident alien among Hittites, his family lacks a self-evident burial site, and he therefore purchases a plot. Sarah's death and burial creates an emotional link to a particular place, but it is an arbitrary one that stands in tension with other, de-territorialized forms of affiliation. While this tension is key to the biblical story, a settler stance attempts to create consistency and certainty in the face of the ambiguity it exposes. Any diasporic elements are made to disappear and then replaced by an emphasis on full ownership of this particular piece of property. Hebron and the edifice Meʿarat ha-Makhpelah are deemed significant by Jewish settlers because they are thought to mark the precise location of the graves of the biblical patriarchs and matriarchs.[2]

Contemporary property rights seem to be directly conferred by the Bible. Yet it is worth noting that many of the passage's key elements regarding Abraham's purchase are erased by this view. I have never heard any resident of Kiryat Arba or the remade Jewish Quarter in Hebron mention, for instance, that Abraham is a resident alien when he carries out the purchase, or that he buries his wife in a Hittite area, or that he bows low in deference to that people when purchasing the property, all features of the Bible that point toward a multilayered diasporic problematic.[3] Abraham may own the property but his ownership is nonetheless underscored by his engagement with other (even pagan-worshipping) people and his outsider status in the area. The settler reading of this passage, in contrast, depends on an interpretative *reframing* that emphasizes his exclusive ownership and control, and this in turn is matched by prevailing social conditions outside the text that grant their particular readings an apparent plausibility. I am therefore interested in the ways that these ideological interpretations become authoritative and the ways secular Jews who support settlement defer to ideological settlers in the name of authenticity. In other words, what social conditions enable this particular application of Jewish tradition to appear both compelling and authentic to settlers and their supporters?

A Settler Tour

To consider the remaking of tradition in lived practice, I turn to the settler tours of Chaim Mageni, a founding resident of Kiryat Arba and guide who with a certain flair expressed key elements of this ideological settler sensibility.[4] During his life, Mageni took hundreds of religious and settler-oriented Jewish tour groups directly into Palestinian neighborhoods throughout the West Bank and stood on location, recreating biblical events in order to bring the past, as he rendered it, to life. His tours consisted of stories, histories, archaeological evidence, hearsay, and hypothetical situations woven together to form a persuasive tale. Though disturbing in many respects, they illustrate how settler claims can seem persuasive for an audience disposed to believe them. To be convincing to the tourists in his charge, in other words, Mageni didn't just quote a biblical passage, but rather, armed with the Bible, he actually sought to reframe his audience's perceptions of an existing Palestinian reality. In short, Mageni's authority as guide and as a devout Jew depended on the facility with which he could invoke and draw on a vast textual tradition that was equally grounded in colonial erasure. Given that most of Mageni's tours took place in Palestinian towns, what sorts of claims seemed to forge attachments to place, shape community, and elicit devotion? These questions remain important because they point to an ideological formation incrementally taking shape in material form. Long before a Jewish settlement actually gets built, settler tours such as these and other practices that remake space lay the groundwork for a material inscription of the biblical past that provides an experience of devout "truth" in the present.

First, let me give a few biographical details about Mageni himself. He grew up in a working-class section of the Bronx and then became active in the Bnei Akiva movement, a religious-oriented Zionist youth movement in the United States that emphasized the importance of immigrating to Israel and working the land. He left the Bronx for Israel just after the 1967 war and studied in Yeshivat Merkaz ha-Rav Kook, the premier national-religious school of higher education that historically has shaped the views of many leaders of the ideological settler movement. Shortly after he finished his education, Mageni was involved in establishing Gush Etzion, the first settlement to be built in the occupied territories south of Jerusalem in the Judean Hills (Mageni Family 2003). Gush Etzion was ideological in the sense that it did not have a direct security rationale attached to it. Rather, its significance was rooted in a national historical memory, reinstating a pre-state settlement that had been overtaken and whose residents were killed in an ambush during the 1948 war. The establishment of this settlement over the Green Line, shortly after the 1967 war and Israeli occupation of the West Bank, took place with the government's blessing. Yet in order to carry it out, religious settlers first tapped into a wider nostalgia for the lost Jewish community that had once lived and died there. The subsequent founding of Kiryat Arba not long afterward employed a different ideological rationale because it was set up for explicitly religious reasons—Hebron was seen by most Israeli Jews as a significant site of origin. Like Gush Etzion, it also was important for the historical Jewish minority that had been massacred there during the 1929 riots that swept Palestine (cf. Mattar 1992:33–49; Cohen 2015). Settlers claimed to be returning to a point of biblical origin and, at the same time, to be renewing a more recent historical Jewish presence, reclaiming either the biblical or lost property of Jews without distinction since both were seen to form a continuous past.

Reconstructing the route of Mageni's tour, I state in my fieldnotes that his bus traveled from Jerusalem through the Palestinian towns of Beit Jala, Bethlehem, Halhoul, and then directly to the city of Hebron. This took place during the early phases of the Oslo period before the Israeli army had actually redeployed from Palestinian city centers and before the Palestinian Authority had assumed control. My notes also document visible signs of the First Intifada, or uprising, in Palestinian areas, specifically garbage cans smoldering and burning tires strewn on the road. Overlooking these features, Mageni stood at the front of the bus with microphone in hand nearly choking on his words. "The road we are on, though paved with asphalt, is one of the oldest arteries known to mankind ... it is the route that our father Abraham, the very first Jew, took from his place of birth in Ur Casdim to the city of Hebron, his hometown within the Land of Israel" (Mageni Family 2003:40). Mageni's tour, in other words, begins by reframing the scene of protest and imbuing settler routes through Palestinian areas with a biblical aura. He tries to convince his audience that the arteries that shape a modern life are not just random asphalt roads secured by the military or avenues of conquest. Rather, they are meaningful as the very roads that have been traversed by Jews throughout their history. He emphasizes that Jews are not strangers in Palestinian areas (and that they are not even settlers) but part of a Jewish presence that has belonged to Hebron from biblical times to the present. Yet, as mentioned, the most extensively documented historical Jewish presence in

Hebron was during Ottoman rule when an Arab-Jewish minority was integrated into a predominantly Muslim society (Klein 2014; Tamari 2009). It was never an armed settler minority allied with a military occupation working to expand the borders of an existing state.

Temporal Linearity as Interpretive Strategy

Mageni's interpretive strategy seems to confirm what Tzvetan Todorov (1999:19), in the context of Columbus's discovery of the New World, referred to as a truth "known in advance."[5] Verifying preexisting points of view is not limited to an ideological settler stance, but in this case it requires erasing or at the very least minimizing the presence of an entire Palestinian residential area marked by a culture, history, and orientation that remains at odds with Mageni's rendering of its primarily Jewish character. Yet Mageni tries to reframe any visible signs of difference in ways that confirm the validity of presumed exclusive Jewish origins and claims. So for instance, while it is true that an Israelite (and proto-Hebraic) presence precedes one that is Aramaic, which in turn predates the Islamic conquests, the spreading of Islam, and the use of Arabic, there are also just as many intersections, continuities, and syntheses that can be pointed out—so much so that a linear time frame emphasizing Jewish origins alone does not do justice to the messy reality at hand.[6]

Place names, in particular, appear to be points where these historical convergences, linguistic resemblances, and intercultural contacts create fields of integration and exchange between Jewish, Muslim, and Christian Arab populations. Yet in the settler imagination, intermingling is recast as linear, and a late-coming Arabic place name always affirms the biblical Hebrew and legitimates contemporary Jewish Israeli claims. The initiate is shown a trajectory that begins with a biblical fact and ends with a modern settlement. Palestinians, when they appear at all, are seen as caretakers and proof of the original truths of the Bible. To provide an example of this, consider the circumlocutions involved in Mageni's discussion of Gilo, a settlement built adjacent to the Palestinian and mainly Christian town of Beit Jala. Gilo today forms part of a population barrier surrounding Palestinian and Bedouin residential areas on the outskirts of Jerusalem. From its inception, it housed Israeli suburban residents in search of affordable housing rather than devout ideological Jews as in Hebron. For this reason, Gilo is considered by most Israelis to be a quality-of-life rather than ideological settlement. Nevertheless, Mageni's narration casts Gilo as part of a biblically inspired regeneration that is taking place throughout "Judea and Samaria," the biblical terms he uses for the West Bank. Gilo becomes a "neighborhood" belonging to the "expanded municipal boundaries of Jerusalem," and it is religiously significant for Jews because it is located near the Tomb of Rachel, an embattled biblical burial site near Bethlehem. Mageni then points to "Arabs," sidestepping the term "Palestinian," and claims they are recent arrivals, having lived on the hills for the past century and a half only, but that even they preserve the evidence of an original biblical Jewish past: He declares that Arabs living on the hill refer to their community by the name "Jala"; "And as we scratch the surface of that Arabic sounding pronunciation, we begin to realize the extent to which local residents in this area, yes, even Arab non-Jewish residents, are preserving the ancient names of the various towns and villages familiar to us from the Tanakh" (Mageni Family 2003:51). By using linguistic resemblances between Hebrew and Arabic and evolutionary linearity, Arabic is deemed useful as a record of all that is biblical and Jewish. Citing for authority chapter 15 (*Pereḳ ṭeṭ-vav*) of the book of Joshua, Mageni explains that *J* and *G* are essentially the same letter and since "Jala" and "Gilo" are cognates, "Jala" confirms the original meaning of "Gilo," which is "to reveal" (ibid). For him, these linguistic resemblances can be used to establish the worthiness of Jewish settler claims to the cultivated fields beside Beit Jala.

Through the collapsing of difference and the renaming of Palestinian places, settlers such as Mageni actively reorient a cognitive field. The visible markers of a Palestinian presence and sites of Christian and Islamic significance are taken to be either surface markers or recent additions that confirm the Jewish past. Difference is collapsed into sameness. The distinct Palestinian history of Beit Jala does not exist apart from a reference to Gilo in Joshua. While scholars have rightly pointed out that Zionism often posits a historical claim based on Jewish origins and appropriates Palestinian culture and history for its own purposes (Masalha 2007; Rose 2005), Mageni's stance endows Israeli national logics with a distinct reading of the Bible and is devoted to a biblical spatial inscription as sacred history. Moreover, an ideological settler's attempt to precisely map a biblical past onto the present evokes not so much land as a general concept as a religious identity forged in relation to a series of distinct locations where biblical events are deemed to have actually occurred.

Claiming Halhoul as a Jewish Site

The Palestinian town of Halhoul on the outskirts of Hebron provides another example of reframing and colonial erasure. In discussing this site, Mageni talks about the ironies of history, and among those he mentions are the following: While actually overlooking Halhoul, he claims to be located at the biblical site of Elonei Mamre (an oak grove) where God appeared to Abraham. Notwithstanding the precision with which he locates the site, there are no oak trees present. Therefore, he quotes the medieval Jewish commentator Rashi to explain that though Abraham pitches a tent and praises god at Elon Moreh and Beit El (biblical places that have now become ideological settlements located outside of the Palestinian cities of Nablus and Ramallah respectively), it is only in Halhoul that Abraham actually settles. Drawing on Rashi for authorization, Mageni again points to a variety of linguistic distinctions in biblical Hebrew that support his claim that, for Jews, Elonei Mamre is actually just as significant as Elon Moreh and Beit El and that by extension Halhoul deserves to be settled on religious grounds. Noting that the Bible describes Abraham's actions with the terms *va-yeshev*, which comes from the Hebrew root "to sit" and *va-yishkon*, which comes from the root "to dwell," Mageni emphasizes that the biblical passage specifically avoids using the term *va-ye'ehal*, which would indicate that Abraham was merely pitching a tent for a temporary period (Mageni Family 2003:74). For him, these linguistic distinctions make up part of the body of historical evidence that Abraham's presence in this physical site was never intended to be temporary and that by extension a Jewish settlement should be built to commemorate his act of settling.

In this interpretation of the biblical text, a contemporary preoccupation with "settling" gets projected back in time and space, rereading the activities and movements of Abraham with linguistic and spatial precision. Yet for the skeptical observer standing before a Palestinian town, there is no external evidence to indicate much of a correspondence between the invoked biblical passage and his interpretation. Mageni proceeds to read the Bible as a text that can be used to map out exact events in a contemporary landscape: "Abraham lived here for 38 years or more before he purchased a plot to bury his wife in the Me'arat ha-Makhpelah," he continues (ibid., 75). By emphasizing the link between Abraham's presence in Elonei Mamre and that of Me'arat ha-Makhpelah, the Tomb of the Patriarchs in Hebron, Halhoul becomes part of a wider network of significant places that need to be inhabited by devout Jews. Using these exegetical, monumental, syllogistic, and experiential elements, the possibilities for claiming other Palestinian places as significant Jewish sites expand exponentially. The issue is not so much belief in the Bible or whether there was ever a Jewish presence in Hebron but how devout settlers narrowly imagine and represent continuities with the Bible and what its implications are for Jewish practices and ethics in the present.

Mageni takes the miraculous event of God appearing to Abraham as bona fide history, raising the question of the Bible's historicity in new ways. Even though the Bible does contain historical material, its mythical events cannot be subsumed into a straightforward historical narrative, and sorting out one domain from the other in this methodical manner involves many leaps of faith. While there is scattered evidence of a place named Mamre outside of Halhoul, there is nothing that links Abraham's actions to the precise biblical sites Mageni points to because it is not actual history.[7] While archaeological evidence often is used to further substantiate religious claims of this sort, it is actually not science alone that is most convincing to believers in these contexts (cf. Abu El-Haj 2002). Rather, a direct experience of being located at the site where a biblical event is thought to have occurred matters more: "It is here that Abraham receives the visitation of three divine messengers" who deliver the news that his wife Sarah is pregnant, Mageni suggests. Yet "here" has no meaning of its own other than indexing the context of the speaker who has uttered it (cf. Silverstein 1976). It is here where Abraham "circumcises himself, and receives a direct Divine revelation" (*va-yera elav ha-Shem*; lit., "and you will look to God"), which includes not only "the promise of receiving land for his offspring" (*le-zar'akha, eten et ha'arets ha-zot*; lit., "To your children I will give this land") but also the actual boundaries of the Land of Israel (Mageni Family 2003:76). These boundaries are believed to lie "from the river of Egypt to the great river of the Euphrates in the northeast," an area that far exceeds that of the territory of modern Israel (ibid.). The discrepancy between (a generally mythical) biblical past and the present created through a speech act is posed as a tension that needs to be resolved through human activity. Mageni's emphasis on Halhoul as the site where biblically mandated boundaries have first been revealed thus provides the rationale for expansion and accords with ideological settler efforts to remake Israel's national borders into those that have religious significance rather than those that have been arbitrarily determined through wars or armistice lines.[8]

Mageni then tells his tour group that God's promise of the land was sealed through sacrifice at an altar. Land promised, biblical boundaries, and places of sacrifice are read as the most critical elements of the Bible, features of Judaism that settlers hope to reinstate through specific settled sites.[9] He searches for material anchors, an altar perhaps, but there are none apart from other Palestinian villages or population centers in the area including the al-Arroub refugee camp to the north. He then quotes the Roman Jewish historian Josephus (37–100 CE), who notes that there had been an altar visible in his day (Mageni Family 2003:77). The altar, according to Mageni's narrative, stood at the center of a Herodian construction and pointing to the ruins of two walls that are present, he notes with authority, "the bulk of what one can see today is Herodian" (ibid.). Mageni's assertions use a variety of sources for evidence, but they are put together in random and often idiosyncratic ways, forming a bricolage of the visual, textual, linguistic, and hypothetical, all rolled into one. He imparts a sense of mystery and discovery to each of his findings. Yet his quest to map the Bible raises the specter of what actually counts as historical evidence, whether it matters more than direct experience, and how the two work together in ways that seem meaningful for a religious audience.

Referring to the missing altar, Mageni deploys a kind of Talmudic logic (and rabbinic style of deduction) to explain its absence: "Although it is too far-fetched to say definitively that here stood the altar that is referred to in the book of Genesis," he remarks, "all the information that we have, and all the material that we see before our eyes makes it impossible to say that it is definitely not [here]" (Mageni Family 2003:80). He then goes on to highlight the sublime place-based sensibility that is essential to his religious claims: "There is something that strikes us Jews in *Eretz Yisrael* today with a feeling of *ḥerdat ḳodesh* [sanctified trepidation] in recognizing that we have the privilege of being right at the site where the Divine revelation to *Avraham Avinu*, our father Abraham, took place" (ibid.). For those on the tour, not much formal training in biblical accounts, the rigors of rabbinic debate, or interpretive exegesis is required. Rather, a sense of awe and heightened experience prevails. On what grounds is this religious solidarity being elicited? Certainly the tour uses biblical quotation to reframe reality and posit a biblical place. Yet its persuasive nature depends not only on the events being invoked but also on Mageni's ability to push a visible Palestinian presence into the background, deeming it to be nonpresent. Just as difference is collapsed into sameness and put in the service of origins during Mageni's discussion of place names, so too does his sidelining of Palestinian lives create possibilities for a settler's subjective investments in the biblical places he uncovers, and this attachment elicits solidarities among this group of onlookers.

Many accounts of the colonial aspect of Zionism from the pre-state period (1888–1945) to the present as well as other historical settler colonialisms talk about a refusal to see natives as having value (Massad 2006; Makdisi 2010). In these accounts, natives are either viewed as backward, as existing in a different moment in time, or as less technologically adept. In Julie Peteet's (2009:41) scholarship on Palestinian refugees, she corroborates these claims by writing that the question of "native competence" and the issue of who is entitled to own the land became a matter of seeing and not seeing. She notes that early Zionists did not consider Palestinians deserving of the land because of their "backwardness," and so they simply were made to disappear from consideration. Zionism's conception of Jewish renewal required that land be depopulated, Peteet argues, and when the 1948 war led to a mass exodus of Palestinians, reality aligned with imagination (ibid). Yet the colonial and utopian dimensions of this earlier iteration of Zionist settlement seem to have been more directly engaged with issues of difference than Mageni's biblical musings are here, suggesting that social hierarchies in the contemporary context have become more rigidly fixed. More direct engagements between different religious communities in Palestine and more contingency in Zionism's ends required, in other words, more explicit forms of differentiation in colonial thought and (labor) practice.

In contrast, ideological settlers like Mageni use discourses that are often far less preoccupied with an explicit working through of difference between Jews and Palestinians. This work of differentiation already operates at a more macro level, through the presence of the military as well as through the occupation's legal and spatial inequalities. Nevertheless, "difference" is not far removed from a settler's biblical invocations, particularly when they occur on-site and are being used to create links to precise physical locations within an explicit social hierarchy. Physical boundaries, legal gray zones, as well as military restrictions in the occupation, in other words, serve to underscore the terms of difference that then get taken up in devout attachments to place.

More significantly, ideological settlers tend to reframe Palestinian lives through forms of exclusion that amplify the more ordinary sorts of social absences and erasures detailed in Erving Goffman's *Frame Analysis* (1986). In his account of the performativity woven into social life, the "real" often depends on a particular framing or staging. Both the subjective investments that an audience brings to the frame and the power-laden rules of engagement serve to create an exclusive focus and scene of action. Others who may be present and who are actually needed to create a heightened sense of the real become shadow presences. In this manner, the staged drama is not only about a particular set of actions but is also an expression of existing power dynamics. Mageni's claims to Halhoul using biblical events shape a distinct reality. Palestinians may be present at the scene of this biblical re-creation and are even critical to its plausibility, but they remain outside its key focus and concerns, a distinct form of erasure. This biblical recasting of the real during tours, walks, and other forms of trespass in Palestinian areas operates as a communicative act that incrementally gets translated into other spatial practices built into the fabric of the everyday.

Bypassing Difference

The macro or structural dimensions of ideological settling need further elaboration here, and it is important to note that they have changed over time. Mageni and other settlers were afforded a freedom of movement through many Palestinian population centers that subsequently became off limits because they were placed under the Palestinian Authority during the post-Oslo period (1993–2000). Both settler and Palestinian populations formerly used the main north-south artery, Route 60, which led from Jenin through Nablus, Ramallah to Jerusalem, and then down to Bethlehem and Hebron. In the spatial reorganization of the West Bank, however, during the post-Oslo period, a system of bypass roads for exclusive settler and military use alone was built, and these went around all the key Palestinian population centers, blocking any Palestinian access roads that might have linked up with them. This bypass infrastructure cut off Palestinian population areas from one another, while consolidating social connections between settlers and the military through quicker travel times.[10] While the Israeli military saw these new roads as necessary to maintain control of the Occupied Territories after withdrawing from Palestinian population centers, Palestinians viewed them as further evidence of land confiscation and settler expansion. The spatial reorganization was also intensely disliked by ideological settlers because they saw it as a way of restricting their access to biblically significant areas. Whereas in Mageni's tour, the Palestinian presence was erased through reframing and the use of biblical events, the large-scale spatial reorganization and infrastructure that occurred after Oslo made this earlier form of erasure seem to be an enduring reality.

The spatial reordering of the West Bank coincided with separating out realms of authority overseeing Palestinian and Jewish settler populations into areas designated as A, B, and C (Israel Ministry of Foreign Affairs 2015: art. 11). The Palestinian Authority was charged with overseeing the civil and security affairs of Palestinian city centers designated as Area A, accounting for 18 percent of the West Bank, while ruling jointly with the Israeli military in the Palestinian towns, villages, and agricultural areas of Area B where they had civil control, comprising an additional 22 percent of the land. The Israeli military continued to be in charge of Jewish settlement areas designated as Area C in the remaining 60 percent of the land (cf. Hammami and Tamari 2001; Hass 2002). Yet, as with all boundaries, these new jurisdictions (A, B, and C) were not as grounded in distinct social realities as they initially appeared, since Palestinians lived in each of these three different zones, and different rules applied in all of them. In cases where Palestinians happened to live close to Jewish settlements, for instance, they were stranded without the security or municipal services of the Palestinian Authority (cf. Sanders 2013). In Hebron, specifically, where ideological settlement had been established directly within a Palestinian urban area, the "exceptional" classification of authority over the area was termed "H1 and H2." While H1 gave control of most of Hebron to the Palestinian Authority, H2 kept 20 percent of the city, including its historic center, and over thirty thousand Palestinian residents directly under the control of the Israeli military, maintaining the pre-Olso condition of direct military rule (Andoni 1997).

Looking Back: A Soldier's Retrospection

These separate population zones enabled settler certainties to take hold, even as they fueled doubts for some Israeli soldiers who were required to fulfill their military duty in Hebron. A soldier's retrospective look at his service in Hebron offers a sobering perspective on the dynamics of rule that form the backdrop to a settler's religious sensibility. After two years of service in Hebron, Yehuda Shaul became disillusioned and founded Breaking the Silence, a well-known Israeli activist group made up of ex-soldiers devoted to collecting anonymous testimonies of those serving in the occupation.[11] They mainly work to educate pre-army draftees in Israel but do not explicitly promote conscientious objection because of its strong stigma. The army remains one of Israel's core institutions, conscripting (mostly Jewish) citizens, male and female, after high school, for three and two years respectively, and is considered a key avenue of social advancement.

Why explore the narratives of a soldier here? First, it speaks to the longstanding and complex relationship that exists between religious settlers and the military. Kiryat Arba settlers, for instance, initially were based in Hebron's military headquarters after a year of squatting in a Palestinian hotel. This early settler relationship with an occupying military force has in effect been carried through to the present. Moreover, an ideological settler's sense of Palestinian lives transpiring beyond the frame is underscored by military restrictions on Palestinian movement that is enforced by checkpoints, curfews, imprisonment, as well as by restrictions on the growth of Palestinian residential areas subject to housing demolitions. Shaul's disillusionment, skepticism, and concern with the randomness of a soldier's actions stands in stark contrast to the certainty of a settler's devotion and sense of continuity. Juxtaposing these two "tours," then, brings out the kinds of military actions taking place in the background and the contradictions they pose for enabling a devout settler's sense of biblical continuity in the present.[12]

On a tour through Hebron, Shaul's knowledge of military matters, as well as his skepticism, contrasted with the resolute attachments of religious settlers. His views were significant not only for their critical sense of what soldiers actually do during their service but because they signal the unraveling of belief in these military actions among young Israelis charged with carrying them out. As an observant Jew with family links to religious settlements, Shaul did not reject fulfilling his military service out of hand. He noted that he went into the military believing that he could conduct himself in a principled fashion, but gradually realized he could not, and remained very doubtful that it was possible to do so in an occupation. It was only at the end of his service that he began to reexamine his experience: "I had some doubts and questions when I was a soldier but I put those questions aside. When you are a soldier, you push questions aside—comradeship is important for this."

Shaul also mentioned that he was not a pacifist—if military service made Israeli lives better and safer and contributed to the defense of Israel, then he believed it was necessary to fulfill it. Yet if "serving" meant a pointless, useless show of force against a mainly civilian Palestinian population and if there was no exit strategy, he was firmly against fulfilling military obligations. His tour provided a good sense of what, from a military standpoint, this security regime looked like—who could enter, who was barred, the kinds of exits and entrances permitted to some classes of people, and the utilitarian naming of checkpoints.[13] For example, he mentioned, the mathematical precision of Checkpoint 300, which masks the human tragedy created through its towering, winding, and labyrinthine rows of iron that look down on the subject pedestrian. As he spoke, Shaul's narratives seemed slightly disengaged, revealing a person who saw the way the military trained him to see, but who was also committed to undoing military logics—orienting the tourist toward the many contradictions etched into a divided landscape. Traveling through some of the same areas Mageni traversed in his tour earlier on, spatial and political realities had hardened. A concrete separation barrier now made its way throughout much of the West Bank:

> The fact that the barrier doesn't match the Green Line creates some, uh, problems, some weird things, and I think this is one of the examples: on your right, you'll see some Palestinian houses, the outskirts of the Palestinian village of Husan, and because of the barrier, encompassing Gush Etzion [settlement bloc], Husan, Datiou, Wadi Fuki, and Halin, around thirty thousand Palestinians in these villages are surrounded 360 degrees because they are stuck between [the boundaries of] Israel, the barrier, Jerusalem, and Gush Etzion. And their way in or out to their main city is through the so-called humanitarian passage, which is this simple tunnel.

He then pointed out a dirt tunnel that has been dug under the road leading from these villages to Khalda and on to the highway into Bethlehem. His perspective highlighted how seemingly rational military decisions gave way to sheer irrationality, as was evident in the many absurdities he pointed to in the fragmented spatial order. Interspersed with a sleek functional road system were, for instance, many mounds of dirt, cinder blocks, stones, and boulders, as well as gates that blocked the entrances from Palestinian villages onto the main road. He also detailed how soldiers carried out policies of separation: "The way these policies [of separation] were implemented and enforced is basically that you would go and put a bunch of big bricks or stones at the entrance from any village to the main road," he noted. If Mageni's route through Palestinian areas invoked a historical route that pointed to the Bible, infusing it with higher purpose, Shaul emphasized the routine activities soldiers carry out in order to create obstacles for Palestinian movement. Indeed, one of the many ironies of the bypass system as it exists today lies in this pairing of sophisticated and planned engineering with hundreds of blocked secondary roads using makeshift and casually produced methods.

Shaul attributed the military's "policies of separation" to the armed conflict in the second phase of the Intifada, from 2000 to 2002, when there were "a lot of attacks on roads, ambushes, open fire on cars, and on settlers who were driving on the older roads," followed by "a phase of suicide bombing attacks in Israel."[14] Yet he also admitted the absurdity of these separation policies because they forced Palestinians to use "back roads that connected villages to one another off the main highways, which for them became the main highways [requiring them] to travel in parts, taking a taxi to one military roadblock, getting off to cross it by foot and taking another taxi, piece by piece." He was grappling with the implications of having helped to double and even triple travel times to places that were actually very close at hand.

Shaul also alleged that he had been required to sow fear among Palestinians and to actively "*li-yetsor teḥushat nirdafut*" (create the sense of being pursued). The idea was that instilling fear would presumably make Palestinian perpetrators afraid to attack, and he maintained that the military used this strategy in order to compensate for no longer directly administering Palestinian population areas. As a soldier, he said that he was required to engage in random and invasive forms of control: "What does it mean to make your presence felt?" he asked rhetorically. "In Hebron, it means twenty-four hours a day, seven days a week you have patrols; you bump into houses, you start your patrol at ten at night till six in the morning in the Old City, walk in the streets, bump into a house (it's not a house you have intelligence about), wake up the family, men on one side, women on the other side, search the place, you can yourself imagine how it looks." In terms of the roads, Shaul emphasized, the military's "flying" checkpoints also meant long hours of patrolling areas at random that yielded little useful information for enhancing security:

> Your mission is basically an eight-hour jeep patrol from the roundabout at Gush Etzion Junction down to Kiryat Arba on Route 60; this is your area and you just have to drive back and forth for eight hours and you just need to invade five houses, doesn't matter where, doesn't matter how long, do two flying checkpoints, doesn't matter which lane, which side of the road, doesn't matter when. You police for fifteen minutes, you can police for forty-five minutes, it can be one after the other, it can be a flying checkpoint

where you check every car, it can be a flying checkpoint where you don't check any cars, just if you are present on the road, and people drive slower and this is part of creating the feeling that we [the military] are all the time everywhere.

Shaul continued working through the trauma of his military service, while harboring suspicion and distrust of those in authority who sent him to carry out the mission. He alleged that there was no way of being an ethical soldier in the occupation and then mentioned the difficulty of reintegrating into civilian life. In sum, Shaul's observations lead us to see the ways in which these military tactics further enable the agency of settlers and even grant greater plausibility to a settler's ideological convictions.

Damaged Eyes: Seeing on the Periphery

In addition to the military occupation, we find other kinds of direct violence that similarly enable the inscription of a settler's biblical vision to take hold. Palestinian farmers living adjacent to settlements who have directly experienced settler harassment and other forms of destruction offer important perspectives here. The Jaber family, owners of lands that border Kiryat Arba, gained the attention of Israeli and international peace activists as victims of verbal harassment and random acts of destruction perpetrated by a gang of male (adolescent) settlers living on the hills above them. The two brothers of the Jaber family, Atta and Habah, were in frequent contact with the Christian Peacemaker Teams (CPT), a Mennonite peacekeeping presence devoted to nonviolence. The CPT operates in the area in order to protect Palestinian families who are vulnerable to settler violence because they live beyond the boundaries of the Palestinian Authority and away from the reach of Israeli police.[15] During a conversation arranged by the CPT, Atta Jaber talked about how his house had been demolished several times by the Israeli military when he tried to build an addition onto it after having been denied the required building permits. Though he had repeatedly applied for these permits, they had not been granted (as they seldom are), and so he decided to expand his house on his own property without them. Then he spoke of High Court petitions, hospitalizations for injuries he incurred trying to defend his house, and after his house was demolished, attempts at rebuilding it along with Israeli and international peace activists. These demolitions and acts of rebuilding became highly visible media events (see, e.g., Israeli Committee Against House Demolitions 2013). His brother Habah, on the other hand, was less of a public figure and far more focused on being able to continue to farm and simply earn a living. A large portion of his land had been confiscated to build a gas station near the entrance of Kiryat Arba and the remainder of it was cut in two by a bypass road for the exclusive use of settlers and the military. Because of these land confiscations, a constant stream of cars coming down from the settlement ran through his vineyards. He lived a besieged life, defending his family from frequent settler attacks coming mainly from Kiryat Arba residents living on the heights above.

Habah Jaber recounted the following: "Settlers came in the middle of night, and they had a saw and cut sixteen grapevines. My father went to the city, and when he came back, he saw all the grapevines wilted. We don't have any weapons. I cannot fight settlers. If they come, I can't defend myself." In telling this story, he talked about his personal sense of fear and frustration at not being able to protect his family. Another time, religious settler students smashed the windows of his house, partially blinding his oldest daughter, who subsequently needed expensive surgeries to restore her vision. Moreover, his four-year-old daughter was traumatized by seeing the settlers proceed to trample over the garden and destroy all the marigolds. He then pointed out that the soldiers stationed in the area did not intervene when settlers attacked his property and that the police did not have the will or manpower to enforce the law. In the face of future acts of settler harassment, Habah considered his choices to be leaving or resisting, and he chose to resist by simply staying put.

Ideological settlers do not see their acts as inherently violent. If they are caught up in direct violence, they claim, it is only in cases of retribution. Moreover, settlers rarely concede the gap between how they imagine the biblical landscape and the uses of force necessary to implement their vision. This gap is characteristic of many other utopian visions that use violence to achieve a desired end. Yet as the experience of Habah Jaber and others Palestinian farmers show, every act of building, renewing, and reenvisioning entails uses of force or direct violence against those resisting that vision by merely living on their land. One portion of the Jaber family land had already been taken over in order to build an expansion, or "neighborhood," of Kiryat Arba known as Harsina (Mount Sinai) years earlier. But the more recent land confiscation in order to put up a gas station left Habah Jaber with a palpable sense of sadness. He alluded to the military decision to confiscate land for what had been termed "public" use: "I talked with the captain. He didn't care about me. He didn't care about my papers [meaning the deed to the land passed down from his grandfather and great grandfather]. He didn't care about the trees. The captain said something I cannot forget: if you have power, you can do whatever you like." Habah Jaber also mentioned that over a hundred of his olive trees had been cut down and carted away.[16] "They didn't give money; they just took the land. I wouldn't have taken gold for it. This is my whole life. How would I exchange it for gold? They offered two shekels (approximately seventy-five cents) for a tree, which was twenty-five years old." He emphasized that none of the Jaber family took money for it: "We would not take that money. Money disappears when you go to the market." Jaber was dismayed that after years of cultivating his land, it could be rendered barren in an instant. "If someone came to photograph it with a camera," he stated, "he would only have seen the soil, stones, and broken pipes for the water."

In contrast to the Jabers' relation to the land, settlers' attachments to sacred sites operate at a much greater remove. Palestinians farmers are invested in Hebron's land because their families have lived on it for generations and it provides them with a basic livelihood. Their attachments to land entail a vast knowledge of soil, water, trees, and agricultural techniques, as well as a mastery of Hebron's long tradition of cultivating grapes and olives. A settler's attachments to place, on the other hand, depend not on the productive capacity of land itself but on bounding it off, fencing it in, and directly residing in it, limiting its agricultural potential in order to sustain a distinct devotional lifestyle.[17]

Moral Hierarchy and Biblical Sites

Reconsidering once again aspects of Mageni's tour in light of the two individual points of view sketched here, those of an Israeli soldier and a Palestinian farmer, it is worth noting that Mageni invoked Jewish mysticism to explain the sorts of vegetation growing on what he considered to be the sacred land beside Bethlehem. His invocation of Judaism's mystical tradition came at the expense of the labor that Palestinians regularly used to cultivate their land. The Land of Israel is "uniquely blessed with seven species" (*erets ḥiṭah u-seʿorah ve-gefen*), replete with two kinds of grain (wheat and barley), as well as five fruits (figs, dates, olives, grapes, and pomegranates) (Mageni Family

2003:43). Quoting the Zohar, a text of Judaism's mystical tradition, Mageni then noted: "It is interesting that in the Zohar, Rav Shimon bar Yochai refers to the special reason that these particular seven species, rather than others, are the ones through which the glory of the Land of Israel is reflected." Though other types of vegetation may be present, they "tend to grow in the valleys [*ba-shefelah*]," and quoting the Aramaic for further effect: "They strive for lowness [*de-hainu sho'afim le-shiflut*] and are not notable" (ibid., 47). He then pointed out that the seven species that grow on the tops of the mountain ranges strive for highness (*sho'afim le-hitromemut*), and are a treasured feature of the Judean Hills to this very day (ibid.). In Mageni's invocation of the Zohar, hilltop vegetation reflected, by its very presence, a moral striving that was directly visible in the morphology of plants. Yet by pointing to rocky areas that could not be cultivated, this mystical tradition also had significant legal implications. The Ottoman land laws, which form part of the body of law used in the West Bank to make decisions on land ownership, equate cultivation with ownership. Land that has been left fallow can be reclaimed by the state for public use after three years (cf. Shehadeh 1988). Incorporated into Israeli military regulations, this law has been used as one of the main rationales for confiscating land on the hilltops and handing it over to ideological settlers, reclassifying Palestinian private property as state land. Mageni's attribution of moral excellence to hilltop vegetation, then, in effect rationalizes the ongoing confiscation of hilltop land where Jewish settlers are given full control.

What of the presence of Bethlehem's Palestinian population? The fields beyond Bethlehem were significant for Mageni because they gave rise to a royal Jewish lineage. It is precisely here, he suggested, that Naomi and her daughter-in-law Ruth approached Bethlehem from the east, and the location where Ruth married Boaz, who bore a son Oved, who in turn had a son Yishai who begot David (Mageni Family 2003:48). Yet it is particularly ironic that Mageni invokes genealogical affiliations such as these to claim land, because they are not easily made to stand in for situated identities. They signal mobile ways of locating the self that can be remembered and recalled by reference to lines of descent rather than through precise geographic links. In the decade after these pronouncements, Bethlehem, a Palestinian city of approximately 25,000 residents, has been made to disappear by another sort of erasure. One section of the bypass system, known as the "tunnels road," transports Jewish settlers and the military directly under the city though a long tunnel. Riding in this tunnel conveys a new sense of normalcy—namely, ideological settlers riding through it have the feeling of being on a direct route home, without any recognition that this Palestinian city has been erased from view.

Conclusion

In sum, as a particular interpretive framing of the biblical "purchase" and a textual rendering of the biblical landscape shows, a settler's religious attachments to Hebron are produced by multiple and complex forms of remaking. In part, this remaking entails the reorientation of an exilic religious tradition so that it always points toward sacred sites, inscriptions in the landscape, and expansions into Palestinian-populated areas. This allows a locally produced settler experience of the Bible as well as other canonical Jewish texts to align with an emerging material reality. Creating these correspondences depends on both narrowing interpretive possibilities and giving biblical passages a material form that brings the past to life in a particular way. This settler experience of the "real" inevitably depends on minimizing or marginalizing a Palestinian presence and on using aspects of Jewish tradition to provide an ethical overlay for this ideological project of erasure. Harassment, trespass, and violence combine with forms of devotion in ways that are no longer deemed antithetical to Jewish authenticity. Whereas in the colonial imagination, "others" were seen as backward and not deserving of resources, this religious framing is more difficult to grapple with because of its authoritative and "authentic" rather than overtly constructed character. While in Goffman, a given frame signaled a kind of social hierarchy that could be unveiled and critiqued, here Palestinian nonrecognition has become part of a more firmly entrenched devotional structure that is becoming more difficult to dismantle even from within its own terms.

Chapter 2

Between Legality and Illegality

In April 1968, a group of young religious students together with a number of families who were disciples of the hawkish leader Rabbi Zvi Yehuda Kook obtained a forty-two-hour permit from Uzi Narkiss, the commanding general of the Israeli Central Command, allowing them to enter the occupied city of Hebron. It was less than a year after the beginning of the occupation, and a permit was needed because the West Bank was closed off to Israeli civilians for overnight stays. This group of religious right activists rented out the Palestinian Park Hotel from its owner Fahd Kawasmeh under the guise of celebrating the Passover holiday in Hebron.[1] They used the week-long time frame of the holiday to initiate a permanent return to a city deemed sacred in the Bible and second only to Jerusalem in its significance for Jews. Once inside Hebron, they squatted in the hotel for approximately six weeks and refused to leave. The Israeli government responded by deferring any binding decision on whether these settlers could remain permanently, relocating them instead to Hebron's military headquarters. Settlers continued to live in this military compound for three years on a putatively temporary basis with their families, until they were granted the right to build the settlement of Kiryat Arba on the city's outskirts.

The government's decision to relocate the Hebron settlers to the military compound was controversial from the outset. It was spurred on in part by a seeming concern with international law, particularly the Fourth Geneva Convention, which prohibits deporting protected persons or introducing part of an occupier's own civilian population into an occupied territory. Under the Geneva Convention, then, an occupying power is bound to maintain the demographic composition of the territory under its control. The decision to transfer settlers into what was considered a de facto temporary military location was

Figure 5. Archival photo of Kiryat Arba (1971). Central Zionist Archives, PHG/1066154.

presented as a compromise that initially positioned these religious devotees as an extension of the occupying military presence. Nominally satisfying this noncivilian and temporary requisite established by the Geneva Conventions, however, gave rise to two important precedents that continue to have deep ramifications: retroactively recognizing as legal settlers' illegal acts of squatting; and giving settlers arms so that the military would not be charged with directly defending them, claiming that they posed an additional security burden.

Legal Indeterminacy and Religious Devotion

This chapter, based on interviews, primary sources, and secondary histories, looks back on the beginning of Hebron's settler presence from the vantage of Kiryat Arba for the lessons on illegality it provides. It does not pretend to be a complete history or a comprehensive study of the legal and illegal elements of the occupation as a whole. Rather, my aim is to highlight the conditions and contradictions that allowed a particular version of Jewish settler practice to be mobilized and take hold. In doing so, I reexamine the idea that historical continuity and the sacredness of Hebron alone led these religious settlers to inhabit the city. Rather, settling Hebron required incrementally producing a site on the ground in a Palestinian city that could then be apprehended as Jewish in order to make settler appeals to origins more plausible. Moreover, legal indeterminacy was the condition that enabled these settler activists to use small tactics on the ground to create an authoritative religious realm. They invoked Jewish law with the aim of diminishing the significance of the Palestinian presence in Hebron because the legal gray zone of the occupation rendered both legal limits and geographic boundaries malleable.[2] This allowed settlers to mobilize limited interpretations of the Bible as prescribing religious imperatives in ways that would not have been possible in a more stable legal environment, granting religious devotion the power to play a key role in confronting the legal authority of the state.

In terms of this legal gray zone and the environment in which an ideological settler project first took hold, it is important to note that the Israeli government's ambiguous position on international law was shaped by the absence of a Palestinian state. Israel contended, in other words, that it had not taken over territory from any "legitimate" sovereign power precisely because it deemed Jordanian rule of Palestinian West Bank areas unlawful. Rather, seeing itself as a liberal occupier, the state claimed that the Geneva Conventions on occupied territory did not apply because it was acting as a temporary trustee, administering the area on behalf of the Palestinian residents living there until they were prepared to rule themselves.

Hebron Settlers as an Inconvenience

According to Shlomo Gazit, who at the time of the Park Hotel takeover was serving as the head of the Unit for the Coordination of Operations in the Territories in the Israeli Labor government, the initial settler presence in Hebron was considered a headache (rather than a serious dilemma) because it afforded no strategic value for the military administration. Hebron had originally been excluded from Deputy Prime Minister Yigal Allon's settlement plan, which, after the 1967 war, envisioned setting up security settlements directly along the Jordanian border and creating a land barrier by annexing a swath of territory ten to fifteen kilometers wide along the Jordan River along with other lands (cf. Shlaim 2014:274). Allon's plan also called for creating two noncontiguous Palestinian areas, including major cities, which he thought would become either demilitarized autonomous Palestinian zones or areas given back to Jordanian rule. His idea was to include as few Palestinians as possible in the most amount of land that could be added to Israeli territory (ibid.). The government later adopted the Allon Plan as its own, modifying it by allowing a settler presence to be established just outside the heavily populated Palestinian city of Hebron and well beyond any previously envisioned security barrier.

Inside Israeli government circles, there was as yet no consensus around what the 1967 war's large territorial gains meant for the long term (ibid.). While those in the Labor-led ruling coalition took the position that a portion of these lands should be exchanged for a lasting peace, others took a more hawkish view and maintained that holding onto all territorial gains made Israel more secure than it had been within its 1949 armistice lines. In trying to explain the ascendance of ideological settlers prompted by the settler group entering Hebron a year later, scholars point to the euphoric mood in Israel immediately after these vast territorial gains (Sprinzak 1991, 1999; Feige 2009). They allege that government officials and a comparatively secular Israeli public were more favorably disposed to a divinely inspired return to cities (East Jerusalem) and sacred sites (e.g., Tomb of the Patriarchs) deemed important in Judaism that had been under Jordanian control for nineteen years. Hebron, then, was seen as a religious exception to strategic plans for the West Bank during a period in which Jerusalem's holy sites were being incorporated into the Israeli state.

As part of this wider security strategy after the war and in accordance with the Allon Plan, Prime Minister Levi Eshkol first began to establish Nahal (Hebrew acronym for *Noar Halutzi Lochem*, lit. "fighting pioneer youth," a special unit of the Israel Defense Forces) military settlements in the newly expanded border areas, particularly in the Jordan Valley, mixing military outposts with agricultural activities (Gazit 2003:251). The idea behind establishing these Nahal settlements was to create an entrenched military presence while using agriculture to finance it and to allow for the possibility that these military bases would eventually be turned into civilian communities along the Jordanian border. Nahal settlements, then, introduced a way of integrating military aims with civilian elements, which in turn made possible a more permanent form of rule over the territories (ibid.). They also enshrined the government stance of "deciding not to decide," which paradoxically extended relatively permanent forms of control while not committing the government to remain in these areas (Demant 1988; Gazit 2003; Zertal and Eldar 2009). In terms of Hebron's ideological settlers, these Nahal settlements served as one of several justifications for religious right civilian-led initiatives. When religious settlers first entered Hebron in 1968, they framed their actions as similar to the Nahal initiatives already in progress, while invoking an earlier history of Zionist settlement to legitimize their actions. Yet at the same time, they began using elements of religious practice to test legal limits from the start, flouting military restrictions in a context where international law was itself being violated.

One could argue that at the time small-scale settler tactics on the ground meant little and that it was mainly government policy (or indecision) that set the stage for the entrenchment of a more radical settler ethos. Moreover, everyday religious lives shaped by adherence to Jewish law and tradition among ideological activists in this period may seem less significant than the bold statements of key religious leaders like Rabbi Moshe Levinger, Rabbi Hanan Porat, and Rabbi Eliezer Waldman, as well as a range of other Labor supporters. Yet the tactics the original Hebron settlers deployed on the ground are worth revisiting here because they tipped the scales in favor of settlement at a moment when a permanent ideological settler presence in Hebron still hung in the balance. Standing between a distinct shift in older forms of messianism, which scholars like Aviezer Ravitzky (1996) and Gideon Aran (1991, 2013) rightly pinpoint as significant in spearheading the religious sensibility of these ideological settlers, and the broader context of government indecision, there exists a whole range of small-scale practices, either directly illegal or pushing the boundaries of a decidedly ambiguous legality, which religious settlers used to create precedents on the ground and to persuade their supporters of the necessity of allowing them to remain in Hebron. By focusing on the spatial or material character of these activities and the agency of these settler-believers, it becomes possible to investigate how distinct forms of Jewish devotion and belief came to depend more fully on the legal gray zone of the Israeli occupation in its first year and thereafter.

Observing Passover, Skirting Authority

Shlomo Gazit, head administrator of the Occupied Territories during the Park Hotel takeover (and later a dissenting voice within the military establishment), contended in an interview that "chance" was largely responsible for allowing the initial settler presence to take hold in Hebron. He recalled that the two military authorities who would have intervened at the time of their entry were absent. Moshe Dayan, the Israeli minister of defense, had just been hospitalized for serious injuries after an archaeological accident (a wall had collapsed on him), and, as head administrator, he too had been away for the week sitting shiva (mourning) for his father, who had died the day before. "The military people who should have and could have intervened at the time didn't exist," he emphasized. He maintained that the settlers used this window of opportunity to enter Hebron illegally. After this, he insisted, it was the government alone who was to blame for their continued presence because its decisions were imposed on military leaders. That is, military leaders were charged with carrying out government orders from above rather than making policy of their own.

Historians Idith Zertal and Akiva Eldar (2009:24), commenting on the settler's illegal entrance into Hebron, allege that Gazit's claims and others like them amount to a rewriting of history in order to minimize the responsibility of those in authority who made a series of decisions that empowered the settlers.[3] As evidence, they quote Rabbi Moshe Levinger, the radical right rabbi leading the Hebron settlers, who refutes the idea that settlers established themselves illegally: "We never told anyone that we were going only to celebrate Passover. The government authorities knew we wanted to settle ... we didn't want to play tricks. Had they followed us closely, they would have seen that any one going to Hebron with Frigidaires and washing machines wasn't intending a pleasure trip. I don't think that it is respectful of the truth, respectful of the Jewish people, or respectful of Hebron to say that if there are Jews in Hebron, it is because we rebelled and went against the will of the government of Israel" (Zertal and Eldar 2009:457n35).

If military authorities disagreed with the settlers, why did they not enforce their own rules and regulations? Did they not have the authority to do so? Gazit's view of the situation was that the military was charged with overseeing Jewish *visitation* to the Tomb of the Patriarchs, while preventing any permanent Jewish presence to take hold in Hebron. Yet he also maintained that he was not about to resign over a dispute that seemed at the time to be relatively insignificant (Gazit 2003:164). In my conversation with him, Gazit emphasized: "I was in charge of an area with a [Palestinian] population of one million people. That was my population, not the Jewish [settlers], but the local Arab Palestinian population." He spoke of the broad task of administering as one of taking on the responsibility for meeting the basic needs of Palestinians, specifically in Nablus, Hebron, Ramallah, and East Jerusalem—

all areas that had absorbed refugees from the 1948 war. He also noted that as a fighting force, the military was not particularly well suited to the task.[4] In comparison, he noted, this band of young religious settlers rightfully warranted little of his attention; not only were their numbers small, but their views were marginal. Gazit, however, did seem to have a basic sympathy with settler demands to pray at the Tomb of the Patriarchs, because he viewed it as a request entirely in keeping with the national ethos of returning to sacred Jewish places.

Occupying a Legal Gray Zone

A distinctive feature of the Israeli occupation has been the legal overlay granted to its many illegal actions. Palestinians under military rule have brought cases to the Israeli High Court, as a check to military actions, but they have rarely won. Ideological settlers, by contrast, have proven more successful in deploying legal ambiguities for their own advantage. The prevailing legal gray zone of the occupation for them meant capitalizing on multiple legal codes that could be invoked and enforced in an ad hoc way by the military administration controlling Palestinian areas. During the early period, this ambiguity allowed for maneuverability with respect to a variety of religious claims as well. Settlers sought to establish precedents, push against regulations, and expand spatial boundaries using religious rationales. These tactics were being deployed on the ground as debates took place in the Israeli government over whether the Occupied Territories were indeed truly "occupied."

Sparked by Israel's (and international) reluctance to recognize either Jordan or Egypt as the legitimate sovereign powers of the Palestinian territories, Israel maintained that it had liberated rather than occupied these areas. To quote Shlomo Gazit again expressing the Israeli government point of view: "Is the West Bank occupied, and if it is, do all the international laws apply to it? If so, we are not allowed to confiscate territory, and we are not allowed to move Israeli civilians into that area. Or is it a *liberated* area, and a traditional homeland of the Jews?" As an indication of the seriousness of this legal predicament, Major General Meir Shamgar, who later became president of the Israeli High Court, advocated using the more neutral-sounding terminology of "held" or "administered" territories (*shetaḥim muḥzakim*) rather than "occupied," striking a compromise among divided Israeli opinion. Disputing the idea that the area had any legitimate sovereign ruler gave Israel greater liberty to reinterpret international law, overriding injunctions against selling or transferring property, while making permanent changes to the area's demographic and Palestinian character.

Though administering Palestinians within the boundaries of the Israeli state had already been established policy by the time of the occupation (cf. Robinson 2013), Gazit alleged that West Bank Palestinians were viewed as a distinct case by military authorities because they were slated to remain beyond the borders of the state. Unlike imposing military rule within its own territory, he contended, where Israel was dealing with Palestinians who were de facto citizens but whose rights had been suspended and would one day be returned, Palestinians in the West Bank were always viewed as a foreign population.[5] Gazit also asserted that because the rights of Palestinian citizens would eventually have to be restored, the military administration necessarily observed more built-in limits in their case than in the West Bank. He did not, however, stipulate what these limits were or address rationales for the indefinite occupation of the West Bank.

Apart from sidestepping international law, the occupying military also shaped its actions by using several inherited legal codes—Ottoman, British, and Jordanian. In Hebron as in other occupied areas, these were merged with Israeli administrative law and new military ordinances further expanding the legal gray zone. The convergence of multiple legal frameworks and codes changed the tenor of many of the original laws, allowing administrators and government officials to pick and choose the laws they saw as most relevant, while confiscating Palestinian land by citing pressing security concerns or military needs.[6] Most prominent among these were the Ottoman-derived laws distinguishing "private" land from that which was "public," or state-owned. In the view of the Israeli military, any lands that were deemed to be state lands, namely, Palestinian lands that had not been cultivated, presumably in keeping with Ottoman precedent, could be taken over to facilitate military rule on behalf of Palestinian civilians. Yet the "public" use of this state land translated into settler and military use alone, to the exclusion of Palestinian needs (Zertal and Eldar 2009:368–72). While confiscated land enabled settler religious claims to take further hold, it also gave rise to settler struggles with the military over whether they, as civilians, would be subject to its authority.

Zertal and Eldar (2009), who contemplate the central role of illegality in their examination of Jewish settlement, hypothesize that the rampant illegality of settlers had its roots in the British Mandate period, where quickly made land claims under the rubric of "stockade and tower" settlements and the sidestepping of British colonial authority were tactics used in laying the foundation for the Jewish state. They do not grapple, however, with the implications of ideological uses of religion in the context of the occupation. Did settler claims based on the Bible contribute to the ways in which those in the government and the military were willing to override law in the name of "tradition"? The record shows that investments in Judaism alone did not directly lead to tolerating settler illegality. Initially, the predominant concern was with not wanting to publicly air internal disputes with the settlers that might jeopardize the military's ability to rule over Palestinians. Moreover, there was also a strong reluctance on the part of the government to deploy the Israeli military against what was perceived to be the nation's devout Jewish representatives.

What is critical here is not, as some imagine, a total consensus across time and space by those on the left and right around settling, but the ability of settlers to influence government decisions with the help of key ministers and supporters, shaping its overall policy without scrutiny or debate. Part of the settlers' growing power, then, coincides with the government's deferred decisions around territorial concessions. As others have noted, the government's unwillingness to commit to a clear position so as to either keep multiple options on the table or to maximize its territorial holdings (Zertal and Eldar 2009; Demant 1988) meant that there was ample opportunity to push through and promote territorial agendas that were not officially sanctioned. Government officials at times turned to settlers as proxies to realize their agendas, but it was equally true that ideological settlers actively cultivated sympathetic officials to establish arenas of their own influence (often based on Jewish observance) beyond the government's immediate control.

The Kiosk Incident

Nothing demonstrates the tactical character of settler actions in testing the limits of the law more clearly than the "kiosk incident" of 1968, a dispute that escalated into a vigorous debate within the Knesset. After having been relocated to Hebron's military headquarters, settlers immediately made arrangements to celebrate the wedding of Benny Katzover, a religious leader who would later become a major figure in the Gush Emunim, the religious movement devoted to establishing ideological settlements in the midseventies. His wedding was poised to be a highly visible affair, and invitations were sent out to a thousand people, including ministers in the government deliberating over whether settlers should be allowed to remain in Hebron permanently. Through this wedding, settlers created an occasion for a large-scale if temporary convergence on the occupied Palestinian city, showcasing the desirability of a Jewish return to Hebron. Settlers then set up a refreshment stand for the wedding guests arriving, and the military duly ignored this seemingly innocuous fact on the ground. However, when settlers pushed for this kiosk to remain standing for future Jewish visitors coming to pray at the Tomb of the Patriarchs, the military governor ordered them to take it down. His view was that settlers could live in military headquarters but not establish new precedents that might influence the government's final decision on Hebron. When they refused, the military governor issued a decree ordering the kiosk's demolition and expelling the settlers who had violated his orders. The incident shows that early on, small-scale settler transgressions such as these had the potential to raise complicated issues regarding whether settlers were subject to military authority, what the occupation's ultimate aims and Israel's territorial ambitions were, and the international cost of expanding Israel's final borders. These small acts on the ground also had political effects that granted the average Hebron settler a direct line to members of the Knesset.

The Knesset debate that took place over the kiosk incident began with the declaration by Yitzhak Raphael of Mafdal (National Religious Party) of a chronology pointing to the continuous Jewish presence in Hebron.[7] His recitation is based on a broad collection of sources from antiquity to the Ottoman period—King David (1000 BCE), Simon Bar Giora (66–70 CE), Eusebius (312 CE), Hieronymus (347–420 CE), Antonius of Piacenza (600 CE), Heraclius (Persian wars, 622–26 CE), Rabbi Ovadia of Barenura (fifteenth century), and Rabbi Gershon of Kitov (approximately 1701–61 CE), among others. In it, he emphasized the continuity of the Jewish presence in Hebron while downplaying the city's Christian and Islamic periods and the modest presence or complete absence of Jews in these periods so as to bolster the settlers' claims. Raphael, like the settlers he supported, argued that the most significant break in this continuous Jewish presence occurred during the 1929 riots in which sixty-seven Hebron Jews were massacred. He noted, then, that the city had recently been liberated from the hand of a "foreign subjugator" (*ha-kovesh ha-zar*), the Jordanian army. Given this long Jewish history, he argued, it was only "natural" (*ṭivʻi*) that when Hebron was recaptured by

Israelis in 1967, the city's Jewish presence would be restored. He mentioned that "tens of thousands" of Jews came to Hebron and found its former Jewish quarter occupied by uncaring residents with its graves desecrated by "fiends" (*zadim*) and that settlers were merely "renewing" historic Jewish connections. Moreover, they did this at great expense to themselves, "living in horrible conditions" (*bi-tena'e diyur ayumim*), first in the rooms of an Arab hotel and later in the rooms of a military compound (Knesset Proceedings 1968:3318).

Raphael emphasized that since the settlers intended to "live in peace and solidarity" (*penehem le-shalom ule-re'ut*) with their Arab neighbors, they were justified in their claims to Hebron, which he compared to the right of a Jew to live in Jerusalem (ibid.). Yet in fact, ideological settlers moving into Hebron developed a deeply adversarial relationship with Palestinians in the city from early on, and it was mainly the perceived legitimacy of their actions by virtue of religious ties that led these ministers to underplay the extent of this conflict. Raphael intoned that he was against expelling the "kiosk four," settlers accused of establishing facts on the ground, unless the heads of the military administration believed, along with the minister of defense, that a Jew dwelling in Hebron was not doing so "justly, rightfully" (*be-din, bi-zekhut*) and that the Jewish government was doing him a favor in allowing him to live there temporarily (ibid., 3319). Shmuel Tamir of the Free Center, a secular right-wing party, sided with Raphael and further accused the Labor government of "implementing a Jewish 'white book' policy [*mediniyut ha-sefer ha-lavan*] like the British," referring to the 1939 limits the British Mandate imposed on Jewish immigration and areas of residence in Palestine (ibid., 3320). The 1939 decision was intended to stem the flow of Jewish refugees entering Palestine from Europe in the wake of Palestinian revolts but was understood by Jews as reneging on a prior commitment to establish a Jewish homeland. By making this comparison, Tamir implied that it was ironic that a Jewish government would battle against British Mandate rule and then present its own obstacles to settlement—though he underplayed the fact that the historical circumstances were now entirely different and that the question of final borders as well as the reality of a military occupation loomed large. Nor did he see supporting ideological settlers as a break with the historical past in which Jews had lived in Hebron as a religious minority under Islamic rule.

By restoring the Jewish presence in this historically significant city for Jews, government supporters of the settlers argued, they were correcting an injustice and returning to Jewish control property that had been unjustly seized, a form of collective restitution. Religious settlers still claim to act in the interest of the former 1929 Jewish community, even though many of its direct descendants reject these claims (cf. Campos 2007). As with any reparations, and even more so in the case of Hebron, the settler claim of "restoring" a Jewish presence leaves aside the question of which historical wrongs need to be addressed, while also overlooking the extent to which Palestinians have been expelled from areas that became Israel. Returning to the Knesset debates again, though, one finds ministers voicing more narrowly construed concerns with justice, emphasizing in this case what they believe ought to be an appropriate punishment for the kiosk four's illegal acts. Linked to this was a brief discussion of the implications of settler illegality for democratic rule within Israel, but only as a secondary concern.

Objecting to what she called the military's proposed "deportation" (*le-haglot*, or "exile") of the so-called kiosk four settlers who refused to take down the kiosk, Esther Raziel-Naor, a member of Gahal (precursor to the Likud), deliberated over the most appropriate punishment. Shouldn't the kiosk, she asked, simply be demolished or did the settlers who built it have to be permanently kicked out of Hebron as the military demanded? Invoking the Israeli military policy of demolition aimed at Palestinians, a policy instituted early on in the occupation (recognized internationally as a form of collective punishment and a human rights violation), Raziel-Naor asserted that demolition alone should be a sufficient punishment because it already included "damage, sorrow, and shame" (*nezek, tsaʿar, u-vushah*). In making a strong case on behalf of settlers, Raziel-Naor sought to compare the treatment of settlers to that of Palestinians and to argue that in this case the settlers were being treated far more harshly. Her impassioned rhetoric ignored the degree to which settlers as citizens occupied a privileged position. Moreover, she did not consider the demolition of Palestinian homes as presenting key ethical dilemmas in its own right. Rather, she reiterated her concern with maintaining firm military control and suggested that punishing settlers would send the wrong signal to Palestinians. Palestinians, she argued, would take the decision to expel the settlers as evidence that the military administration was weak and internally divided (Knesset Proceedings 1968:3321).

The left-leaning side of the Knesset debate showcased the minority opinion of Uri Avnery of Ha-ʿOlam ha-Zeh–Koaḥ Hadash, the former editor of the anti-establishment newsweekly *Ha-ʿOlam ha-Zeh*, which lent its name to the first radical protest party elected to the Knesset. Avnery highlighted the issue of "democracy" and the "force of law" as the overriding issue highlighted by the kiosk incident. He did not seem to object to the military occupation per se at the time but rather to the fact that the settlers' actions were bypassing the protocols of democratic debate in the Knesset by establishing facts on the ground. Debate by elected officials alone, he argued, should determine whether Hebron ought to be included in the final borders of the state. He also added that the settler actions had bypassed the authority of the military governor and that this governor had been given the responsibility to maintain the safety of settlers along with that of soldiers, sons of "the people," by none other than the elected government itself. Finally, Avnery asked rhetorically: "Is this military governor [sic] not worthy of receiving the full backing of the House, of this people, and especially of those who are the first to cheer when the name of the Israel Defense Forces is mentioned?" He also maintained that the devout beliefs of ideological settlers meant little in view of their provocative actions: "It doesn't matter what they believe in," Avnery asserted, "what matters is what they did" (ibid., 3322).

In trying to square religious claims with illegal actions, Avnery argued that ideological settlers came to Hebron "via fraud and deceit," which he maintained was motivated by an antidemocratic wish to "force the cabinet to take a decision not wanted by the people." Avnery then posed a stark choice from the outset—either tolerate the lawless tendencies of these settlers in the name of keeping a Jewish presence in Hebron or remove these violators and emphasize "democracy" as a principle that takes precedence over all religious claims. He argued that just as within Israel, "democracy must override Jewish claims," so too in the occupation, the "rule of law" must prevail over religious interests. He then stated that the settlers were provoking the civilian Palestinian population with "armed demonstrations and parades," while eroding any trust that existed between the military governor and Muhammad Ali al-Ja'abari, the mayor of Hebron (ibid., 3323). In the end, the government did override the military decision to expel the Hebron settlers.

Settler actions against Palestinians from their earliest days in Hebron included harassment, property damage, and land takeover. This lack of constraint further led settlers to take the law into their own hands and engage in acts of violence. Tobias Kelly (2006, 2009) highlights the significance of uneven applications of the law and the uncertainty it brings for generating everyday violence. Moreover, the absence of fixed borders added to this state of lawlessness.[8] The issue of national borders and, along with it, the border-making aspects of settlement resonate with a final set of issues that came to light in these early Knesset debates. The position of Knesset member Tawfik Toubi (New Communist List) was particularly resonant. As the only Palestinian in government to weigh in on the Hebron settler affair, he asked the following: If Israel's expanded national borders based on religious claims meant that Israel might never "normalize" its status and might fail to be more fully integrated in the Middle East, was the Jewish settler claim to Hebron worth preserving? Since the Arab world stood in support of Palestinian aspirations for self-rule and against colonial legacies, he asserted, would not the presence of Jewish settlers jeopardize Israel's standing in the region and inflame anti-Jewish sentiment? He then maintained that Jewish settlers would impede any long-term peace proposal and thwart wider diplomatic efforts between Israel and the Arab world: "Peace and annexation, coexistence and occupation—these are opposites" (ibid., 3326). This was a prescient warning.

How did the settlers use their religious expertise and invocations of Jewish origins to establish their own authoritative sphere in light of these debates? Even though a number of ministers in the Israeli Knesset supported settlers and argued for Hebron's religious relevance to Jews, they did not dictate the terms of religious observance on the ground. The problem of deciding how Judaism would be invoked in the contemporary context of Hebron was essentially left to settler activists to determine for themselves. A careful look at settler activities, claims, and practices beyond the kiosk incident reveals that they were adapting different elements of Judaism to the military context, accentuating aspects of tradition that would better assure their permanence in Hebron. In most cases, the spatial dimensions of Jewish religious laws were put to new use, aligning with the aims of prolonging military rule. The military, in turn, also took on the role of facilitating certain religious affairs outside of its area of expertise, including building a *mikve* (ritual purification bath) within its headquarters. Gazit remarked on these activities: "In my eight years of political life, the one and only *mikve* I was responsible for building was there [in military headquarters]. Because they said we need a *mikve*, so I had to build a *mikve*. The question is not whether I liked it or not—once they were there and I was responsible for them from an administrative point of view, I had to look after their needs." He intimated that he would have preferred that this religious settler group not be housed in the military's headquarters. Overseeing a *mikve* or deciding whether or not settlers should directly "pray before the tomb of Yitzhak" was not the sort of thing that he, as a military man, had been trained to do. Yet through these demands to accommodate aspects of Jewish observance, as well as political appeals and protests on behalf of religious needs, ideological settlers tested the limits of military authority and at times sought to redirect the course of the government's plans. Religious invocations and practices then became structuring moments in their own the right.

A Settler Memoir

When religious settlers first moved to Hebron's military compound, they sought to establish themselves as a civilian presence within it. Chaim Simons's (2003) memoir of the first three years in Hebron is written from a nostalgic-triumphalist point of view. He states that one of his first concerns was to create, along with other members of his vanguard group (the "nucleus," or *garin*), a distinct mode of transportation from Jerusalem to Hebron. Settlers deemed it necessary to establish a separate Jewish form of transport rather than using the Palestinian taxi services or military vehicles that existed at the time. While transportation was their focus, the venture also was a way of creating a route to a more permanent Jewish settler presence in the West Bank before any settlement had been built. Their first application for a taxi license was turned down, however, because Hebron fell outside of the jurisdiction of Israel's Ministry of Transport and because these licenses were presumably not issued for closed military zones. Yet as a result of applying several times, the Hebron settlers were placed under the housing authority of East Jerusalem, an Israeli government branch that dealt with administrative matters in an area "outside of Israel proper" that would be later annexed. When finally granted the license, the settlers were reclassified as civilians living beyond the nation's boundaries, a distinct status that had previously applied only to residents of East Jerusalem.

Next, settlers tackled the issue of voting—they posed the question of whether they would have to travel to Jerusalem to vote or if not whether these services would be extended to their residence in military headquarters. If the latter, was their voting booth considered to be "civilian," and were they exercising their rights as citizens, or was it a military amenity they were entitled to by virtue of living in a military base? Through these small tactics, settlers sought to define a distinct status for themselves—pushing to categorize themselves as "civilian" in order to sustain a devout religious community beyond existing Israeli borders. They then sought to extend an array of other municipal services (electricity, telephone lines, housing) to their living quarters in Hebron in order to enable themselves to be more permanently situated—all means of blurring the 1967 boundary.

While applying for these permissions, settlers also began to set out the parameters for a distinct form of religious observance. Simons notes that he and two other fellow Hebron settlers, together with others living in Kfar Etzion, enrolled in a course on ritual slaughter (*sheḥiṭah*) offered by the Israeli Ministry of Religious Affairs. Though the course was not intended for anyone residing beyond Israel's boundaries, they enrolled anyway, sharpening blades in a metal bunker while drawing a government stipend to further their religious education. In effect, Jewish dietary law (*kashrut*) was being practiced and adapted to a military context. Predicated on extending prescribed forms of humane treatment to animals, the settler version of kashrut instead entailed carefully raising and killing chickens but cleaning them with underpaid (Palestinian) child labor used to pluck feathers. Jewish dietary law was being observed, but the ethical basis of its humane practices shifted away from its universal application to particular aims. With newly acquired skills in ritual slaughter, for instance, Simons recalls that his settler group decided to begin hatching chickens in another bunker for a ready supply of kosher meat. The bunker, however, was located in an area that was also being used for target practice (Simons 2003:40), and Simons voices his concern that the competition for space meant that a person shooting might injure the chickens (ibid., 44). The colonial backdrop of the whole enterprise, however, was taken for granted even though it depended on the direct exploitation of labor arrangements that the occupation afforded. Simons asserts, for instance, that he initially sought to hire adult "Arab" workers for this venture but instead recruited several children because they were cheaper. His observations and narrations create a strange blend of concern and indifference, focusing variously on not inflicting pain on animals, while ignoring the plight of Palestinians under occupation.

Media and publicity also were being used in ways that were not traditionally part of even Modern Orthodox Jewish observance, though they became important dimensions of this evolving ideological settler sphere of influence. They were, from the very beginning, a way of taking the Hebron settler cause directly to the public and amplifying a settler presence by minimizing their small numbers. Ideological settlers used media coverage early on, then as now, to generate debate around each of their actions and legal infractions. They also continued to spread their ideas through religious education. A yeshiva for higher religious learning was among the first institutions to be established in

Hebron's military headquarters and the earliest in the West Bank. Headed by the radical rabbis Dov Lior and Eliezer Waldman, it too was approved by the government as a temporary presence and as a way of meeting the educational needs of Hebron's religious male students. The yeshiva was then moved to Kiryat Arba and served as a model for the hybrid *hesder yeshivot* that followed, a distinctly religious and military blend of educational institutions built within other ideological settlements that were established later. These yeshivot taught a form of observant Judaism that was made to seem entirely compatible with military service, further distinguishing the inner circles of the settler movement from other sectors of Jewish Orthodoxy.

Likud's Rise to Power

Following Israel's near defeat during the 1973 Yom Kippur War, the right-wing Likud Party won the 1977 elections for the first time in Israeli history. Likud's hawkish platform under Prime Minister Menachim Begin proclaimed the right of the Jewish people to settle anywhere in the West Bank and Gaza. Aligning itself with these changing agendas, the Gush Emunim (religious protest movement) reached new heights and mobilized its followers around charismatic rabbis, increasing its settlement activity in the West Bank exponentially. The Likud platform was partly shaped by ongoing peace negotiations with Egypt and was intended to placate Israeli dissent over returning the Rafah Salient and the Sinai Peninsula in exchange for a peace treaty (Gazit 2003). During these negotiations, Likud was also concerned with suppressing Palestinian nationalism and preventing Palestinian alliances with Syria, Libya, and Tunisia, countries that backed Palestinian opposition to Egypt's recognition of Israel.

Yet while ideological settlers and the Likud were aligned on ethnic matters, differences remained over tolerance for religious observance. Likud's brand of nationalism was secular, and it viewed the settlers' claims with a degree of suspicion, preferring to advocate a Jewish right to the land based on "blood and soil" tendencies alone; Herut, the Likud Party's precursor, had espoused these ideas even before its rise to power (Taub 2011). Ariel Sharon, whom Begin appointed as minister of agriculture, as well as minister without portfolio of the Settlement Committee, and whom many view as the most prominent advocate of settlement in the government during this period, was more inclined to use right-wing ethnonationalism as a way of implementing his own political vision (Zertal and Eldar 2009:56).[9] He approved the building of dozens of settlements directly adjacent to Palestinian towns and villages, giving them a more explicit role in surveillance over these areas.

Matti Drobles and Official Approval

As a way of looking more closely at the collaborations and tensions between nationalism and religion under Likud, it is worth considering the manner in which Matti Drobles, a former Knesset member (1972–77), characterized the religious character of Hebron's settlers. Drobles was an old-style planner and Herut functionary who served as head of the Settlement Department of the Jewish Agency during the early 1980s. His recollections point to a blind spot with respect to the forms of religious radicalization taking place under his watch. He was first put in charge of planning for the Jewish Agency, specifically because the Jewish Agency and the World Zionist Organization (WZO), technically international and nongovernmental organizations, took over the responsibility for West Bank settlement in this period as a way of masking the Israeli government's involvement.

Heading the Settlement Department, Drobles was officially made responsible for establishing new communities within all of Israel, but under the direction of Sharon, he funneled most of the department's resources into building up the West Bank and Gaza to the exclusion of other areas within Israel's borders (Zertal and Eldar 2009). In an interview, Drobles recounted: "I established a program to settle 100,000 Jews in the West Bank—and everyone laughed, particularly the left. They thought that it was all a fantasy." He recalled presenting Sharon with his plans and easily getting the stamp of approval for all the settlements he proposed, stressing that it was done with the "appropriate permits," because otherwise it would not have been "legal." His emphasis on getting legal authorization is indicative of a shift in approach, so that in contrast to that of the earlier Labor regime, Sharon's plans began to take on the aura of established "official, legal, and government-planned activity" (ibid., 58).

Drobles went on to emphasize that the "right" to settle over the Green Line came out of nationalist rather than religious imperatives. Explaining this further, he pointed out that he and others in Herut had never entirely accepted the partition of Palestine and that, like the Revisionist leader Zeev Jabotinsky (1880–1940) whom he admired, he believed Israel's territory should include "both banks of the Jordan." In this manner, Drobles gave renewed credence to an idea that had long fallen out of fashion with the establishment of the Israeli state. He also rejected diplomatic efforts that would have continued to trade land for peace, even though he admitted it created the "conundrum" of what to do with the Palestinians living on these lands. Drobles like others in the Likud did not believe Palestinians could be integrated into Israel, but unlike the radical settlers he in effect supported, he did not agree with their call to transfer Palestinians out of the area. His position was contradictory because he also dismissed Palestinian aspirations for statehood and (mis)characterized the Palestinian population as "Bedouin," meaning nomadic people.[10] Ideological settlers, on the other hand, seemed to him more like secular Zionists who had ties to the land and built kibbutzim and moshavim (agricultural settlements) that extended "beyond the walls of religious cities." As a consequence, he rejected any specific term referring to a settler over the Green Line, favoring "Zionist" instead: "A settler beyond the Green Line (*mitnaḥel*) is a person who comes to conquer an area that is not his . . . for me, there is no such thing."[11]

Not only did conquest not exist for Drobles, but he maintained that a distinctly "religious" motivation played no significant role in settling whatsoever: "We didn't decide to settle for religious reasons [*dat*] but for *historical* ones." This purely "historical" claim meant, in his view, that Hebron settlers were replacing Jewish communities, which had previously lived in Hebron (the 1929 community), and it was only by *coincidence* that the Jews who established the new settler presence in Hebron "happened to be religious." Drobles's tendency to frame ideological claims as "nonreligious" apparently came out of his concern with not allowing religious views to prevail. It was enough, he maintained, to allow those who "happen to be observant" to live as they preferred among others with a similar lifestyle rather than as a minority in a secular context. He then noted that "ideological" settlers had been very useful for leading the move into "undesirable" areas: "Those who come out of ideology give up and do everything. Everything stems from it, all the development—the roads, the electricity, water, homes, books, and universities. . . . But *oy va voy* [God forbid] if a state is built *just* on 'ideology.' We will create an Iranian state." Members of the Likud, then, saw their alliance with Hebron settlers as tactical, and a settler religious "ideology" as the "glue or cement" of expansion and development. Drobles added that he never supported religious fanaticism but sought to settle only *normal* people who were concerned for Israel and who followed its laws.

Between Legality and Illegality 69

Yet in point of fact, this emphasis on "normalcy," echoing earlier Zionist claims, masks notable cases of settler violence linked to Kiryat Arba and Hebron that took place during this period. For instance, the militant rabbi Meir Kahane's radical group Kach had by then established a following, and Kiryat Arba was where Kahane himself had lived.[12] In addition, a vigilante offshoot of the Gush Emunim, whose key members also were based in Kiryat Arba, carried out a series of attacks targeting Palestinians in Hebron in retaliation for earlier attacks on settlers. As a result of these terror attacks, Palestinian mayors and students were maimed and killed. Further, a new underground terror cell, the Jewish Underground (ha-Maḥteret ha-Yehudit), was amassing weaponry to bomb the Dome of the Rock in Jerusalem when its members, some of whom also lived in Kiryat Arba, were arrested and the mosque's destruction was narrowly averted. In sum, a shift had taken place from earlier forms of illegality using small-scale incursions to more direct violence planned and perpetrated by religious right radicals who were prepared to take the law into their own hands in order to institute their vision of change. Incursions testing the boundaries of legality did not disappear entirely however, but instead took on a distinctly gendered form. As we see in the next chapter, it was Kiryat Arba's religious women and children who were charged with spearheading a potentially volatile act of settler expansion, moving directly into Palestinian Hebron's Old City. They squatted there for over a year, using extensive media coverage while domesticating the space in ways that mobilized Jewish motherhood for settler gains.

Chapter 3

Motherhood and Property Takeover

Entering Kiryat Arba, one is immediately struck by the vast number of children playing in basic but well-tended playgrounds. The sheer numbers of children at play outdoors, tended by their mothers in flowing skirts and stylish hats, form a panoply of nature and nurture that disrupts one's preconceptions of how militant religious communities are constituted. At the same time, many volatile Jewish settler incursions, property takeovers, and expansions into Palestinian space have featured women and children as key actors. In the midst of dozens of children, women's use of force in taking over property is masked, and they have succeeded in taking control of Palestinian areas where men have failed. In participating in protests, religious women have made ample use of gendered logics to forge distinct opportunities for territorial acquisition, as well as to enhance settler control of confiscated property. The maternalist activism of ideological settler women directed against the military and government is based on the presumption that women's primary identities are linked to motherhood and that the care of children has a natural connection with renewing a Jewish spiritual home in Palestinian Hebron. In this chapter, I show that in the course of taking over property, women in ideological settlements like Kiryat Arba are directly engaged in reworking the gendered dimension of Jewish observance, expanding the boundaries of the Jewish community and its matrilineal basis, while at the same time using the ethos of motherhood to remake the character of Palestinian areas being taken over. Women's activism, in other words, operates not only through the invocation of Jewish origins and continuity but in relation to the ethical imperatives that figure deeply in culturally specific perceptions of religious motherhood. Nurturance, the protection of children, and women's altruistic actions are therefore deployed to forcefully expand the territorial reach of a devout settler presence in the occupation (cf. Neuman 2004c).

If settler tours and narrations as well as the legal gray zone have prepared the ground for property confiscation, here I focus on key historical cases of expansion and takeover from Kiryat Arba into the Old City of Hebron, exploring how the gendered aspects of Jewish observance are being leveraged. I show how women's agency in these takeovers becomes embedded in practices of care and attachment that originate in the domestic sphere (cf. Avishai 2008). This is because women's activism in this context is often predicated on actions that employ a fluid notion of domestic space (transposing bonds between mothers and children onto physical sites) and on the harnessing of women's reproductive capacity to create bodies that can be distributed over newly claimed spaces. In this manner, motherhood not only ensures the continuity of the Jewish settler community through reproduction, following the biblical injunction to "be fruitful and multiply," but also takes on a militant character. It serves to create and socialize a ready population inducted from birth into political activities that give rise to a religious social transformation. As evident from the cases that follow, the stakes of nurturance and tenor of maternal bonds are often instrumentalized for political ends.

Scholars writing on "fundamentalism" have observed that many of these religious movements see their modes of observance as a return to traditional values and that this return often depends on asserting greater patriarchal control over women's reproduction and sexuality (cf. Ruthven 2007; Riesebrodt 1998). They have debated whether women in fundamentalist movements are coerced into participating in overtly patriarchal frameworks or whether women choose to abide by them of their own accord because they find inherent value in them. This debate, in other words, showcases the choices and constraints of women in fundamentalist communities and the degree to which analysts need to take a collective ethos into account when considering issues of individual choice and freedom. In this chapter, I consider aspects of these debates in relation to the ideological settler context, exploring the modalities of women's agency in events of religious right activism featuring the value of motherhood. I then consider a second arena of influence, namely, that of the military-colonial context and the asymmetries propelled by it that shape settler women's lives over and against Palestinian lives.

Religious Women Establish a Stronghold

To illustrate some of the most striking examples of maternalist politics, I focus on three key ideological settler incursions during the 1970s that ultimately resulted in establishing a new permanent settler presence in the Old City of Hebron. The aim of these incursions was to establish a strong link from the settlement of Kiryat Arba to other Palestinian property that settlers considered to be historically significant because of the 1929 Jewish community that had been expelled from Hebron in anticolonial riots during the British Mandate. The events of incursion featuring women and children include holding secret circumcisions in Hebron's Tomb of the Patriarchs, reinstating a Jewish cemetery with the corpse of an infant, and taking over the former Daboya clinic. These acts are significant because they led to a permanent Jewish settler presence that was and continues to be particularly incendiary due to its location directly in urban Palestinian areas. They are also important as telling incidents of maternalist activism that set precedents only a decade after religious settlers entered the occupied West Bank. Moreover, they illustrate how women sought to establish "facts" on the ground that put forth a civilian rather than militarized image of settlement. In all of these cases, women's attempts to expand a settler presence were not sanctioned by the Israeli government, but they eventually led to official recognition and support of new strongholds.

From the beginning of their entry into the Hebron, settlers focused on expanding places of control by placing women and children at the forefront of volatile political protests. They then exploited the run-down and untended condition of many buildings in Palestinian Hebron's Old City to expand beyond Kiryat Arba. In these incursions and permanent takeovers, Sarah Nachshon played a key role as a devout mother and religious settler leader. Embodying the blend of activism and motherhood endorsed by the ideological settler movement and residents of Kiryat Arba, she was seen as a role model by insiders. The first of the events that she led focused on circumcision. Prior to Kiryat Arba's establishment in 1971, when its founding group was still being housed in Hebron's military headquarters, Nachshon delivered her fifth child. Her son's birth was as much a public as a personal affair, and Nachshon recalled receiving a letter from Deputy Prime Minister Yigal Allon congratulating her but declining an invitation to join the Hebron settlers in celebration. According to Nachshon, it was also during this period that the

Knesset considered and rejected a settler petition to allow circumcisions to take place in Meʿarat ha-Makhpelah (Tomb of the Patriarchs).

In 1968 the Tomb of the Patriarchs had been opened by the Israeli military administration to Jewish visitors for what they deemed to be limited worship, effectively changing the Ottoman-era status quo that confined Jewish prayer to the outer steps of the Ibrahimi Mosque. Nachshon's actions sought to consolidate a permanent Jewish settler presence directly inside the mosque. In our conversation, Nachshon recalled that she had illegally circumcised all her sons in Meʿarat ha-Makhpelah—using wine—and that after the circumcision of her sixth son, the newspapers had written that a "most secretive mission" (*mivtsaʿ sodi be-yoter*) had been completed. However, when she circumcised her seventh son there as well, the police raided the ceremony:

> A police officer arrived in the middle of the circumcision, arrested my husband, and confiscated the bottle of wine, saying, "you'll be doing time for this" [*ʿal zeh atah tashev*] because it is forbidden to bring wine into the Meʿarat ha-Makhpelah. We continued celebrating the mitzvah [circumcision] and suddenly my husband returned and joined us. So everyone asked, how did you get released so quickly? And he answered, "due to lack of evidence" [*mi-ḥoser hokhahot*]. The person who performed the circumcision was still around but the wine bottle had disappeared.

Nachshon relished the final twist of this story, emphasizing the incompetence of the police who arrested her husband but inadvertently drank the wine themselves before it could be used as the material evidence necessary to prosecute him in court. The laxity of law enforcement due to a "lack of evidence" has been prevalent throughout the occupation, but as early as the 1980s it sparked Israeli dissent. In this period, fourteen Israeli law professors lodged complaints over irregularities in law enforcement against the abuses of Jewish settlers, particularly in Hebron. An investigation headed by Yehudit Karp, Israel's deputy attorney general, known as the Karp Commission, discovered a pervasive failure to apply the law in cases of settler threats, harassment, acts of aggression, and violence due to "lack of evidence" (Karp 1984). Yet the report's warnings went unheeded, and Karp ultimately resigned from the commission in protest. Nachshon, on the other hand, celebrated her victory over the law. She also expressed her respect for soldiers rather than the police stationed in Hebron in a manner that is indicative of the way settlers often exploit divided arenas of control between the police and military to facilitate incursions into Palestinian areas.

Relative to the transgressions, abuses, and threats that have given the community of Kiryat Arba its notoriety, the efforts of Nachshon to circumcise her sons in the Tomb of the Patriarchs struck me as relatively innocuous in the early part of my fieldwork. Yet as a pattern of confrontational standoffs and takeovers led by women emerged, I came to see its broader implications for settler land claims. First, her actions desecrated Muslim sacred space with wine, but they also remade an element of Jewish ritual. The circumcision of a newborn, which ordinarily takes place on the eighth day of life became joined to a place-specific location. That is, the temporal dimension of circumcision (and the marking of the body) was given a new spatial emphasis, specifically linking it to the site where the ritual took place. Circumcision of the male body, a symbol of affiliation with a people, instead depended on being carried out in a site of Jewish origins near graves.[1] These realignments between the body, people, and place underwrite the attachments to renewed biblical sites that religious settlers champion as a key component of their religious identity. Nachshon's actions in turn opened the way for other settlers to occupy, use, and change the character of the Ibrahimi Mosque until a defining area of it (including the courtyard) was designated as a synagogue.

While the Israeli government was a party to the idea that this mosque was built over an important site in Jewish history, aligning government and settler interests, Jewish origins were not in themselves expressed as the key rationale for partitioning the site. Rather, ongoing conflicts between Jewish settlers and Palestinian residents amounted to, in the eyes of military authorities, an administrative problem, and the military saw itself as helping to maintain law and order, mediating conflicts that they assumed had been fostered from as far back as the British Mandate. The Israeli government's 1975 decision to divide the mosque drew heavily on ideas of partition, which sought to settle conflict and violence through the division and reapportioning of space. It eventually assigned Jews and Muslims to separate spaces of worship but did not address the underlying causes of the conflict. Nachshon's activism alone, then, did not deliver half the Ibrahimi Mosque into the hands of Jewish settlers, but it helped pave the way for creating a permanent Jewish presence at the site. It also emboldened settlers to expand into other places through gendered practices that played on both strategic uses of emotion and the historical-nostalgic significance of a lost Jewish community.

Remaking Hebron's Old Jewish Cemetery

In 1975 Sarah Nachshon again led a push to gain control of Palestinian land in an area that settlers referred to as the Jewish cemetery. The cemetery of Hebron's historical Jewish community was no longer in use after the British relocated Hebron's Jews to Jerusalem in 1929 following attacks against them and again during the boycotts of 1936. Under Jordanian rule, the cemetery had become a cultivated field near the Old City. Stone fragments with Hebrew discovered in the field provided the basis for settlers to argue that a new Jewish cemetery needed to be built at that precise location. With great symbolic ceremony, they reinscribed the names of the Jewish victims of the 1929 riots on new tombstones and resumed burials of recently deceased settlers, placing them beside those of the 1929 victims to formalize the idea of Jewish return and continuity, incorporating former graves and the tombstones of these victims into new settler claims.[2]

The 1975 standoff between the Israel Defense Forces (IDF) and Sarah Nachshon together with other Kiryat Arba religious students over the burial of her son had been mentioned to me in other contexts as an illustration of the activism of religious settler women, but Nachshon herself recounted the event as a larger-than-life feat in which she was the primary heroine. I later found other versions of this story, contributing to its status as a founding myth among settler women and their supporters. Nachshon's version was as follows: After returning from a hospital in Jerusalem with her deceased baby, she initiated a standoff with the military. She did not mention, however, as other versions do, that she was joined by several hundred religious (male) yeshiva students who were prepared to support the standoff on her behalf. Learning beforehand of her intent to bury the child in an area then off-limits to settlers, Nachshon recounted that the military blocked the road and refused to let her car pass: "The military stopped me and said that they had orders not to let me bury my son in Hebron, and that I should either return immediately [lahzor hazarah] to Jerusalem or bury him in an area near the industrial zone inside the boundaries of Kiryat Arba."

It had been standard practice in the early seventies to bury Kiryat Arba's dead in Jerusalem, but settlers wanted to establish a Jewish cemetery in Hebron because they thought that it would make their relatively new presence harder to withdraw. As a result, Nachshon refused the military's compromise of establishing the cemetery on land that had already been designated as belonging to the settlement. She recalled that, when she asked the soldiers why she could not bury her son (who had died of crib death) in a "historic place" for Jews, they replied that it was because Hebron would be returned to the "Arabs." As a way of expressing her determination that Hebron be kept in Jewish hands Nachshon took action:

> I waited by the Glass Junction at the entrance of Kiryat Arba, and it was full of military, and they didn't let me pass. I stood there until it was beginning to get dark and so I said, "Well, you won't let me pass on account of this baby who hasn't even sinned ['od lo ḥaṭa] so I'll take him out [of the car] and I'll begin to walk by foot." When they saw that I really took the [deceased] baby out, they ran to their walkie-talkies [in order to contact to their superiors]. I began to walk down to Hebron, to the cemetery, and when they saw me walking, they ran after me. They told the minister of defense that they couldn't stop me, that he should send other soldiers as reinforcements because they alone couldn't do it. And then the minister relented and said, "Okay, let her pass" [tenu lah la'avor].

In this confrontation with the military and the minister of defense (Shimon Peres at the time), Nachshon claims a right to bury her seventh child in Hebron because he had lived and died there, if only briefly. She stages grief and resolve to advocate laying her son to rest in his spiritual home. Her display of emotion does a certain gendering work in this militarized and male-dominated context (cf. Helman 1999). Moreover as a mother she appears to be engaged in an act of civil disobedience rather than illegality, breaking the law only to uphold religious values. She presents herself as a valiant fighter, essentially ignoring the fact that her agency depends on legal ambiguity in an occupied zone. Her actions pit not only motherhood against the military but also yeshiva students against the soldiers charged with protecting them, dividing the loyalties of religious men who will subsequently serve in the military.

While it is impossible to know precisely why the soldiers did not intervene, it is clear that Nachshon's staging of herself as a distraught yet defiant mother was persuasive because of the cultural value accorded to the religious and maternal roles she inhabited. Her demonstration of grief challenged the "masculine" pragmatism and emotional restraint required of soldiers.[3] Her maternal loss seemed to tie into the idea of personal sacrifice in what the anthropologist Eyal Ben-Ari (2001:112) sees as one of the overarching myths of "participation in war." Her protest made use of this trope of sacrifice familiar in Israeli nationalism, merging religious right activism and private emotional realms together. At the same time, she invoked the Talmud to justify her demand. With this religious overlay, then, she confounded received understandings of the political—as mainly secular and masculine—placing the intimacy of a devout mother at the forefront of a campaign to establish a territorial stronghold.

Nachshon linked her actions to divine will. She recalled in conversation that night had fallen by the time she and a number of Kiryat Arba residents arrived at the place that settlers had designated to be Kiryat Arba's cemetery. In darkness, she dramatically spoke to those gathered:

> It is told in the Gemara [elucidation of the Mishnah in the Talmud] that a father and son went off to journey and the son was tired and his father put him up on his shoulders and the boy asked him: "Father, when will we arrive in the big city?" The father replied: "The moment you see the cemetery [lit. house of life] you will know that we are close to the city because near every city there is a cemetery." So, if today we opened the old cemetery which is in Hebron, we will also succeed in creating a big city here. And second, history is like a circle. Abraham bought the cave for his wife Sarah. And my name is Sarah and I've now acquired a burial site for my son Abraham after three thousand years.

When considering how Nachshon invokes Gemara, there are at least two interpretive shifts that seem to authorize the renewal of the Jewish cemetery in a Palestinian area. The first is the move away from ethical and philosophical musings in Judaism on temporality, human frailty, and mortality, toward a renewed emphasis on territoriality as a value in its own right. The second is the process of highlighting death as a rationale for permanent habitation. To provide a clearer sense of this reworking, let us again consider Nachshon's uses of tradition. She quotes the Midrash (a didactic tale preserved with the written version of Jewish oral law) of father and son approaching a city. The tale has been linked to a passage that appears in Psalms, which seems to have little to do with the way Nachshon deploys it: "May the Lord answer you in time of trouble" (Tehillim [Psalms] 20:2). The tale conveys the significance of maintaining trust in God even when one is tired and loses hope. For this reason, the father replies: "My son, take the following as a sign [of things to be]: If you see a cemetery, know that the city is not far away" (Midrash Tehillim 20). Religious scholars have interpreted the tale as messianic, considering the cemetery to mark the outer limit of human history. Yet in Nachshon's reworking of it, opening a Jewish cemetery at a particular site in itself seems to be a core value. Place replaces time in this inverted settler view, and death signals possibilities for the presence of human life. As in other settler interpretations, Nachshon's invocation sees death as the beginning of renewal, giving it new value. She also asserts that as "Sarah" she is following a pattern set by Abraham's burial. Repetition and reiteration, then, become a key emphasis in the settler ethos of renewal here.

After Nachshon recounted this incident, I discovered that a version of it had been published in a 1979 article written for the Israel newspaper *Haaretz* by Amos Elon, who was long considered a leading liberal journalist. In this article, Elon expresses his discomfort with what he sees as the "fanaticism" of settlers, invoking a liberal Israeli point of view. His article contends that Israeli authorities are being manipulated and held captive by the actions of a small group of religious extremists, and with regard to Nachshon's activism, he writes: "Ever since I heard this story, with all its violent morbidity, from her husband himself, I can't get it out of my mind. What dark guilt feeling resided within this unfortunate woman who turned the corpse of her child into a political flag? What was she compensating? What was she trying to prove to herself and others? It is inconceivable that only 'political motives' played a role in this drama. There is something raw, barbaric, fearsome in it, reminiscent of Greek mythology. This woman, like Antigone, will stop at nothing" (Elon 1997:74). Elon sees Nachshon as a modern-day Antigone, like the sister who relentlessly attempts to bury her brother in spite of the fact that the ruler has forbidden it. Antigone's fatal flaw is that she pursues her own form of justice, and relentlessly tries to secure her brother's honor until it destroys both her and her supporters who end up committing suicide. There is a certain truth to the idea that Nachshon is forging her own path by invoking the Talmud as the ethical basis for her actions. Yet Elon overemphasizes individual affect and religious devotion to the exclusion of the key social and cultural values within ideological settlements that shape these maternalist protests. Moreover, he sidesteps the fault lines of Israeli politics and policy that legitimate them. Like other liberal Israeli observers, then, Elon psychologizes Nachshon, attributing to her the passion of all devout believers. Absent is any consideration of this staging of motherhood and strategic uses of religion in the wider context of the military occupation.

Women and Children's Takeover of the Daboya Clinic

This story of Nachshon's burial shows how elements of Judaism have been given a greater territorial and site-specific aspect in the maternalist activism of religious settlers. As a place taken over by a religious mother, it has been staged for political ends, seeming to stand in contrast to threats of Palestinian violence, as well as to uses of military force in the area. This staging was further elaborated in the 1979 takeover of the Daboya building, again involving Nachshon, but this time working together with other high-profile settler women activists and their children. This was the first of a series of buildings that were taken over in the Old City of Hebron beyond the boundaries of Kiryat Arba. In light of a number of failed attempts to secure the building by men, a *Ma'ariv* newspaper account notes: "15 mothers and their 35 children drove a truck down from Kiryat Arba at night, loaded with mattresses, cooking burners, gas containers, water, a refrigerator, laundry lines, and a chemical toilet."[4] Women and children squatted in the run-down building for well over a year, forcing the government to choose between letting them remain or using soldiers to evict them. The takeover of the Daboya building, which settlers call Beit Hadassah after the Jewish clinic that operated there prior to 1929, was considered to be a historic achievement and a symbol of women's leadership in actively shaping settler aims within the patriarchal strictures of this local religious and political community. In recalling the takeover, Nachshon explained: "The government didn't know what to do with us. They asked us to leave and we answered that we wouldn't clear out willingly [*lo nitpaneh me-ratson*]. We stayed day after day without windows and without doors—everything wrecked [*paruts*] and destroyed [*harus*]—without running water, without electricity, without anything, but yet we stayed there in order to build and reconstruct the city of Hebron."

The camping out of women and children in this building for over a year coincided with Israeli peace negotiations with Egypt. It appears that government indecision was more expedient than ousting these women by force in the internally divided atmosphere of the Camp David Accords, when Israel agreed to return the Sinai Peninsula in exchange for the normalization of relations with Egypt (Quandt 1986). Settler women capitalized on this moment, forcing the state to "balance out" the politically unpopular decision of having to evacuate the Sinai with a more resolute commitment to controlling the West Bank. Further, a 1980 Palestinian attack on a group of yeshiva (religious) students joining the women for Sabbath celebrations left six dead, catalyzing a final decision to allow a more permanent settler presence to be established at the site where the women had squatted. The building, renamed Beit Hadassah, was remodeled with public funding, and fifty armed families were allowed to move in under military guard. It became the first heavily fortified building of what settlers refer to as the "Jewish Quarter," which is located directly in the Old City of Palestinian Hebron.

A number of striking tactics were used to reconstitute the Jewish domestic character of Palestinian space during this confrontation. Mothers were presented to the media as protecting their vulnerable children, while claiming a right to return to an area that had housed Jews until the 1929 riots. They lay claim to the vacant ground floor, which under Jordanian rule had served as a United Nations Relief and Works Agency (UNRWA) school for Palestinian refugees. The tumultuous transformations of the intervening half century after 1929—the establishment of the Israeli state and expulsion of the Palestinian population from its territory, Jordanian rule over the West Bank and the Israeli military occupation after the 1967 war—were obfuscated by this plea for the return of lost Jewish property. By their very presence, the women attempted to remake the materiality of the building, and their staging of domesticity signaled the benevolent side of a colonial project. They showcased an idealized view of Jewish family life struggling to maintain itself in difficult

Figure 6. A British soldier standing guard in front of the original Beit Hadassah clinic in the wake of the 1929 riots. Central Zionist Archives, PHG/1016875.

Figure 7. The fortified Beit Hadassah enclave in Hebron's Old City, 2005. It also houses a museum devoted to Hebron's Jewish history and a memorial to the victims of the 1929 massacre. Photo by author.

living conditions. Expansion of settler control became enmeshed with caring for children, and creating a home tacitly played on a sense of domesticating a place considered wild and destroyed.

While women claimed the Daboya complex on behalf of their children, they were also securing releases from the military to go to the hospital to give birth and returning to the site with their newborn infants. They put their children in vulnerable positions and used them as buffers in the ensuing tensions.[5] Though settler women seized and lay claim to a Palestinian area, their staging of maternalism neutralized the illegality of their actions and attempted to persuade the Israeli public of the integrity of their intent if not their deeds. The exceptionalism of religious mothers in the political arena, in other words, allowed these women to appear as nonpolitical actors whose demands were rooted in authentic emotion and domestic needs rather than explicit political goals.[6]

The left-wing *Haaretz* reporter Yehuda Litani was skeptical of what he

Figure 8. Settler women activists from Kiryat Arba and their children in the Daboya takeover, 1979. The women include Miriam Levinger (back row, on the right, seated apart), Geula Cohen (seated to the left of Levinger, hair uncovered), and Sarah Nachshon (second row, on the left, between the two women wearing glasses). From Miriam Levinger, "'Hadassah Women'—Hebron Style," *Ve-Shavu Banim le-Gvulam* (pamphlet, Beit Hadassah no. 1).

understood to be the Gush Emunim's "ploys" and demands made on behalf of these women and children.[7] Its board of male leaders who championed their cause insisted that the government "take immediate measures to stop confining [*le-hasarat ha-matsor*] the women of Beit Hadassah and to fulfill its promise of renewing the Jewish settlement of Hebron" (Litani 1979). The Gush Emunim also opposed all settlements in which Jews were "confined behind barbed wire fences" and argued that their move into city centers was aimed at expanding the narrow limits the government had imposed on them (ibid.). One might add to Litani's observations that using domesticity as a means of seizing property was ironically represented by these Gush leaders as an example of women's confinement and vulnerability, further erasing their political agency.

Litani notes a pattern in settler attempts to reach an agreement with the government and spells out the critical stages that lead a takeover to be retroactively recognized and reinstated as a permanent presence.

In the first stage, the Gush activists established a conciliatory dialogue with their supporters in the Government ([Ariel] Sharon, [Zevulun] Hammer, [Haim] Landau) and in the Knesset. They emphasized before their supporters that the insufferable situation [*matsav bilti nisbal*]—in conditions not even worthy of being called conditions [*tena'im lo tena'im*]—in which women and children are living in Beit Hadassah cannot continue. Having convinced those already convinced, Sharon and Hammer then took the opportunity during a vulnerable point in the Israeli negotiations with Egypt to propose "at this stage" to allow a limited number of families from Kiryat Arba to bring their belongings to the place. It stands to reason that some ministers, primarily Deputy Prime Minister Yigal Yadin, Minister of Defense Ezer Weizmann, and Minister of Foreign Affairs Moshe Dayan, objected but the matter hinges on one vote [*yitkabel 'al ḥudo shel ḳol*]—whether to allow 10 families from Kiryat Arba to remain for a "three month period until discussion is resumed on the matter" [*'ad le-diyun meḥudash ba-noś'e*] of Beit Hadassah and a number of other buildings in the center of Hebron that belonged to Jews. (Litani 1979).

The important elements of this protest that Litani highlights include the demand for improving the living conditions of women and children, negotiations with sympathetic politicians, and a willingness to defer decisions on the permanent status of settlers. In part, the claims made by Gush Emunim on behalf of women and children are possible because they are not perceived to be political actors in their own right. Men become their primary advocates in bureaucratic-administrative circles, tactically negotiating the public-private divide on their behalf. Meanwhile on the ground, women and children are not only continuing to carry out domestic activities but also transforming the materiality of a Palestinian space and remaking relationships of power to go along with it. Seizing property, in this case, also entails the erasure of more heterogeneous periods of the past.

Shortly after settlers moved into Beit Hadassah, a Palestinian tenant whose upholstery business was on the first floor underneath formally lodged a complaint. He came to his shop one morning to find that his ceiling initially had a 1.5-meter hole in it and subsequently that the entire ceiling had been destroyed (Karp 1984). Settlers alleged that dancing during a religious celebration had resulted in the destruction. After several attempts to fix the ceiling, this tenant came to his shop and found that all of its contents had been looted: "On Sunday when the upholsterer returned to his store and saw it empty, without a ceiling but with a stairwell, he sat by the entrance and began to bemoan his bitter fate. Three armed Beit Hadassah men approached him, and asked him to remove himself from the store entrance. When he told them it was his store, the three stepped on his feet, and pushed him outside, where they picked him up and either led or threw him out. The man's son showed up while this was going on. He too was pushed in the incident and his left arm was hurt as a result" (Karp 1984). The Karp Commission report then notes that the Palestinian shopkeeper was pressured to leave because settlers claimed that his stairwell had once led to a synagogue. In short, the takeover of the Daboya building led to more direct forms of settler violence against Palestinian residents. Maternalist protest, then, often masks the actual violence of subsequent takeovers and territorial expansion. The domestic sphere—replete with cooking pans, nursing infants, and playing children—is mobilized to produce a distinct form of politics. It is seen as that which extends the reach of life-sustaining activities, but it often supports rather than disrupts uses of violence as well as the prevailing patriarchal order (ibid.).[8] Moreover, in standoffs with the military, engagements with government sympathizers, and interactions with a broader public, women as disaffected mothers make use of traditional gender roles to reinforce rather than unseat the privileges conferred on them by military rule.

Analyzing Women's Actions

At issue in analyzing many cases of maternalist protest is not only how daily life in the settlement generates and shapes women's agency but also whether women's political activity expands traditional gender roles. Tamar El-Or and Gideon Aran (1995), Israeli anthropologists who have extensively studied the religious settler movement, analyze a well-known case of women's settler activism that established the (initially illegal) outpost of Rachelim near Shilo and conclude that it offers a critique of the intensely ideological and masculine character of settlement. Focusing on the strategies of right-wing women who in 1991 set up a large tent outside the boundaries of Shilo to mark the site where a religious mother of seven had been killed in a Palestinian attack, the authors note that these women succeeded in extending the initial Jewish mourning period several times over until their temporary outpost became a permanent settlement. El-Or and Aran maintain that the women achieved their aims without violence, treating soldiers as their own sons, and that their actions offered an important alternative to the aggressive male ethos that often prevails in settlements. They therefore see settler women activists as having the potential to bridge the stark divide between feminists and fundamentalists even if they are not sympathetic with feminist values.

In response to their analysis, I turn to the gendered dimensions of daily life in Kiryat Arba, which suggest that women's participation in takeovers is a complement to rather than critique of patriarchal values.[9] While religious settler women may temporarily change existing relationships of power within the religious home by taking control of spaces outside of it during protests, they continue to uphold traditional patriarchal values within their own community and a colonial stance with respect to the Palestinian areas they seek to dominate (cf. H. Herzog 2009). Settler women, in other words, may use fluid boundaries as a resource for takeovers, but by doing so they do not readily seek to challenge the place of women in the domestic sphere, nor paradoxically even the classic idea that true political expression requires "freeing oneself from the satisfaction of wants and needs in the household" (Arendt 1998:118).

Within the religious home as well as beyond it in protests, caring for many children tends to reinforce rather than subvert women's traditional religious roles. For instance, emphasizing prolific reproduction over formal study has meant that settler women tend to have less formal religious education (Aviad 1983; El-Or 2002; Rapoport et al. 1995). This has led to debates within religious circles over how much formal education women ought to have and whether child rearing and household activities are sufficient to fulfill a woman's religious obligations. In lieu of religious study, settler women tend to be directly involved in overseeing the lived experience of Judaism within the home, as well as in socializing their children to sustain an ideological project. This life path tends to be shaped more by custom than by choice: arranged marriage, motherhood, and overseeing large families linked to values of purity and modesty. Yet at the same time, religious settler women see their lives as freer and more modern than those of Jewish women who live according to Haredi Judaism. They may hold jobs outside the home and be active within their communities—focusing on social work, teaching, and other arenas of social welfare that strengthen settler aims. They also participate in women's organizations such as Women in Green—which Haredi religious women are barred from doing.

The Eruv as the Expanding Boundary of the Private

Though the households of large religious families tend to regularly flow into the communal spaces of the settlement, this expansion of the home or private sphere is formalized in the weekly observance of the Sabbath. Sabbath in settlements as in other Jewish contexts depends not only on ceasing to labor but on the leveling out of internal social and economic differences within the community. Internal community bonds are strengthened, individual claims are diminished, and, in contrast to the average workday, the designation of what counts as private shifts outward (Neusner 1991, 2001). Closely linked to this shift is a second related principle, which is that a person is required to stay put. Staying put from the vantage of a nomadic or fleeing population is the essence of what it means to be at home (ibid.). In the settlement as elsewhere, an *eruv*, or physical boundary, marks the limits of a Jewish community. The eruv is also significant because it creates a ritually defined and expanded private sphere, allowing a particular form of labor, namely, carrying, to occur within its confines. Though carrying in public is a prohibited form of labor, it is nonetheless permitted within the home and in this ritually defined public space.[10]

In Kiryat Arba, the boundaries of home and community and the sorts of things that can be carried within each on the Sabbath present ideological settlers with religious dilemmas. It is here that one can observe another reorientation of Judaism taking place that is critical to its distinct settler form. Distinctions between public and private spheres are complicated by a settlement's military surroundings, and its cutoff relationship from adjacent Palestinian residential areas. The barbed-wire security fence encircling the settlement is taken to count as an eruv, making it possible for its devout residents to see the entire settlement an extended private sphere during the Sabbath. Much like any other observant Jewish community with an eruv, carrying within a settlement's common areas does not violate a prohibition on carrying, and religious settlers keep keys in their pockets, push baby carriages, and carry food to their neighbors. Matters become more complicated, however, when it comes to carrying weaponry, because that does violate the Sabbath and is allowed only in exceptional circumstances. Settlers in Kiryat Arba mentioned that one could carry a gun on the Sabbath because the entire settlement was encircled with a fence. Soldiers could also carry their weapons in order to guard the settlement. Yet there was no consensus on whether this ritually defined space extended into Palestinian neighborhoods, which settlers regularly traverse in order to worship in the Tomb of the Patriarchs. The path to it leads out from the confines of the settlement and is not included in the eruv. Because settlers did carry weapons, they tended to invoke the halakhic principle of being in "mortal danger" (*pikuaḥ nefesh*). This meant that they were allowed to violate the Sabbath in order to defend their own lives. However, this precedent didn't entirely fit Kiryat Arba's circumstances because "mortal danger" cannot be a situation one willingly enters into, nor is it supposed to be a regular occurrence. It is the exception to the rule, and regularly carrying a gun to the synagogue on Sabbath is therefore a category error. In other words, the regular use and normalization of guns in ideological settlements has shifted the strictures around Sabbath observance, placing greater value on maintaining control and security in an occupation over the idea of equalizing differences, while encouraging movement and expansion over staying put.

Staging Motherhood: Circulations and Gendered Legitimations

If the "staging" of motherhood in ideological settler protest achieves its political aims in a national context by using emotion and empathy, it also seeks to engage international audiences by downplaying the role settler women have in furthering the occupation. As representations of devout Jewish motherhood circulate, the militarized aspect of religious settler lives gets written out of the narrative frame. Maternal relationships with children are emphasized, as is the threat of (Palestinian) violence to a cherished civilian domesticity. June Leavitt, a religious settler and American immigrant to Kiryat Arba, for instance, published her diaries in German and English, highlighting the distinctive elements of her life in Hebron. In *Storm of Terror: A Hebron Mother's Diary*, Leavitt (2002) uses the personal qualities of the diary genre to capture the essence of an anxiety-ridden mother caught in a maelstrom of violence. As an expatriate living in Kiryat Arba, who emigrated to Israel in 1979, Leavitt writes about a personal journey that begins with her own search for spirituality while homesteading in upstate New York and ends up as a Jewish mother living through the perils of life in Hebron. She is critical of the religious community in which she lives and prefers synthetic versions of religiosity that merge heavy doses of new age spirituality with Jewish Orthodoxy. In her self-portrait, she depicts herself as taking Bach flower remedies, practicing yoga, and consulting a bioenergy pendulum to steady her frayed nerves while bullets hail from all directions.[11] The terms "settlers" and "settlements" are only mentioned occasionally, and instead the reader is (dis)oriented by a dizzying assortment of place names and locations that cross boundaries and create a distinct "topography" of violence.

A diary entry dated November 8, 2000, for instance, writes out women's agency in takeovers, while focusing on the mother-child bond alone. In it, she details a surprise visit to her daughter Estie, who at the time is serving as a young soldier stationed at Beit Hadassah, guarding the settler families who live there permanently. Leavitt tells us that her daughter has taken the atypical step of volunteering for military combat and is protecting civilians who live in this imperiled zone. Wracked with worry, Leavitt decides to drive down from her home in Kiryat Arba to her daughter's military post five minutes away on an unpaved Palestinian road "intended for donkeys and carts" (Leavitt 2002:37). In her discussion of Beit Hadassah she makes no mention of the 1979 takeover of the building by Kiryat Arba's women and children. Rather, her narrative is as follows:

> Beit Hadassah, a three-story apartment building erected for Jewish settlers in the 1980s, took its name from the Hadassah Clinic upon which it is built. The clinic had served the Jewish community until the pogrom of 1929. Mixed in with the smell of medicines, which amazingly still lingers there, are memories. There is the apartment of the [Jewish] pharmacist who, despite his wife's pleas not to, opened the door to his Arab neighbors believing they were truly in need of his help. In cruel and bizarre ways, they tortured, then murdered him and his family. (Ibid., 39)

Leavitt portrays Beit Hadassah as a building "erected for settlers," choosing to minimize its actual takeover. She focuses on the 1929 pogrom against Jews under British rule. This historical context rather than that of 1968 Hebron settlement serves as the background for her narrations of motherhood disrupted by the ravages of war. She portrays her daughter Estie as a child soldier dressed in oversize fatigues, holding a rifle that exceeds her height, while standing alone at night in a treacherous place. When Leavitt shows up unannounced, Estie's words are: "Are you crazy, why did you come down here?" Leavitt answers, "How can I let you be in the crossfire while I sit quietly at home?" (ibid., 38–39). While the incident actually shows the close links of domestic life in Kiryat Arba to Israeli soldiers and settler soldiers deployed within Palestinian areas under occupation, Leavitt's diary represents only the universal challenges of motherhood—letting a daughter grow up and become an adult in the face of danger.

Later in the book, Leavitt talks about how she rushes, under a barrage of bullets, to the house of a "militant Zionist" friend and mother of eleven children. Who wouldn't be militant, her diary suggests, if one had to protect eleven children? Here again the trope of motherhood is being used empathically and universally, writing out the distinctiveness of the military context and colonial ethos it participates in. This idealization of women and children underscores the way Leavitt is able to use her own position as a mother to downplay the severity of settler aggression. Within the confines of Anat Cohen's home, Leavitt talks about the lost innocence of her friend's children, while referring to the plight of Jewish children alone.[12] Leavitt (ibid., 40) writes: "I listened to Anat. She told us that the Jewish children of Hebron have become experts in weaponry. They know from the sound of each blast what kind of guns the Arabs are using; the Kalachnikovs we gave them; the M-16s they stole; the heavy machine guns they smuggled in. In the morning, the children collect spent cartridges, not only from the street but also from the floors of their living rooms and bedrooms."

Jewish children living in Hebron are portrayed as vulnerable throughout Leavitt's diary. Bullets whiz through their bedrooms and graze the sleeves of their bathrobes. In her entries, children have lost their innocence by being exposed to war. Yet there is little indication that their families have actively chosen to live in a high-conflict zone for ideological reasons. Also absent in Leavitt's telling is the sense that ideological settler children are not only victims of war but part of a demographic battle being waged by their parents and that they have no choice in the matter. Leavitt simply reiterates her belief that God will intervene on their behalf. Yet her statement seems to be more of an article of faith than fact as we read of the many child amputees and orphans that fill her pages. She also downplays prolific reproduction and the reality that children make up the majority of residents in both this Jewish Quarter and other ideological settlements like Kiryat Arba.

An Ongoing Process

As a postscript to this discussion, I turn to the case of Shalhevet Pas, a ten-month-old girl, the daughter of settlers who lived in Hebron's Jewish Quarter; she was killed by a Palestinian sniper. According to several newspaper accounts, the family refused to bury the infant in order to publicize the demand to reoccupy Abu Sneinah, a Palestinian residential area overlooking the Jewish settler site from which the shot is alleged to have originated.[13] Large banners appeared with the baby's picture over the caption "Give Abu Sneinah back to the IDF. Shalhevet's blood cries: Stop Terror." Baby photos were plastered on bumper stickers, across the walls of Hebron, and worn around the necks of militant settlers. Accounts also state that thousands from Kiryat Arba and surrounding settlements poured into Hebron for the funeral held in the Jewish cemetery, which had been taken over by Nachshon some twenty-eight years earlier. A group of these settlers reportedly used the funeral as an occasion to set shops in the Hebron market on fire and indiscriminately damage Palestinian property. To make matters worse, the Israeli military heavily shelled the area during battles with Palestinian gunmen.

In addition to being a vehicle for territorial expansion, the Shalhevet Pas case illustrates how children in settlements can be used to bridge sympathies beyond a religious settler constituency. Two years after the incident, the Israeli Ministry of Foreign Affairs continued to post a memorial page to the infant on its website.[14] The caption of her photo, which was juxtaposed to the official state seal, stated that the Pas family had endured several tragedies in Hebron's Avraham Avinu neighborhood, and it went on to detail these Palestinian attacks, while omitting significant aspects of the ideological settler context the family lived in. Thus, in the government's show of support for the nation's civilians (particularly women and children) and innocent victims, the boundary between Israel proper and its ideological settler strongholds was erased. Through the death of this baby, there was a shared emphasis on victimization that extended far beyond this settler enclave, blurring the distinctions between religious and secular, radical and mainstream right, as well as right and left in a Jewish claim to Hebron beyond the Green Line. Yet the volatility of this provisional consensus was highlighted by the arrest, July 17, 2003 (and subsequent conviction), of Yitzhak Pas, the infant's father, because of his underground activities and charges of belonging to a terrorist cell (Gutman 2003).[15]

On Prolific Reproduction

Settlements have some of the highest fertility rates in a nation that already has one of the highest fertility rates among Western industrialized societies.[16] In terms of their reproductive decisions, married settler women often navigate between social constraint and personal choice. As the ethnographic record suggests, numbers of children or spacing between births is not entirely a matter of individual preference. This is because women are subject to patriarchal frameworks that take the form of (male) rabbinical interpretations of Jewish law. In religious settlements like Kiryat Arba, the use of birth control is not explicitly forbidden, but it requires consultation with a rabbi, which is an obstacle for women who want fewer children. At the same time, many religious settler women actively embrace the idea of having large families as a way of rejecting secular feminist norms and living in a way that they understand to be an authentic expression of Jewish identity.

The sense of needing to have many children did in fact predetermine the paths and limit the choices of devout settler women. While there seemed to be security gained from meeting social expectations, the lack of control in reproductive matters created internal conflicts. Though women did not often express ambivalence over the obligation to have many children, they did voice some reservations over the issue of spacing between births. Orit, a twenty-five-year-old resident who had recently given birth to her first child, for instance, told me that it would be inappropriate to seek the rabbi's approval for using birth control at such an early stage in establishing her family because he would only approve it for pressing health concerns.[17] She mentioned that having four children was a minimum for obtaining his permission and that neither personal issues nor financial matters warranted choosing longer intervals between births. Like other sleep-deprived parents, she seemed exhausted from caring for her six-month-old and was also having trouble losing the weight she gained during pregnancy. Yet she dismissed these struggles by saying that exhaustion was a woman's lot and that many other women she knew in the settlement complained of being tired too. She also felt isolated. Caring for her baby left her with little time to socialize, and she mentioned that she missed the close circle of women friends she had during high school. She told me that women her age were mainly busy caring for their families and there was hardly any time to get together with them.

The decision to establish a large family conferred status, and there were not many other roles available to women outside of becoming mothers. Women who were divorcées or widows also participated in the maternal ethos of the settlement by providing childcare services or teaching in kindergarten or elementary schools. Adolescent girls, especially the oldest daughters of large families, devoted a considerable share of their time to helping their mothers manage household responsibilities and were therefore socialized into the ethos of motherhood early on.

Yet in spite of these constraints many religious women saw raising large families as a spiritual path that contrasted with the base materialism of secular Israeli society. Devorah, a veritable matriarch in her midfifties—mother of ten, grandmother of dozens, and great-grandmother of two—pondered the significance of women's roles within Kiryat Arba. She expressed the view that having many children showed that Jewish women were holding out against the march of "Western civilization" because recent trends in other communities had demonstrated that the more education women acquired, the fewer children they had. As a woman who was educated and the head of a large family, Devorah felt that she was forging an authentically Jewish way of life. Moreover, she believed that she and other religious women were helping to pave the way for Jewish redemption because each birth increased the possibility that the Messiah would be born.[18] She contrasted these values with those of secular Israelis, whose families were small, she alleged, only because of their greater emphasis on material wants and needs. Children in large religious families, she was quick to point out, were not nearly as materialistic and were content with hand-me-downs or fashions that were not "the latest."

Religious motherhood carried with it the assumption of a ready-made intimate bond between women and children that was not always the case given the many demands of running large households and the financial hardship associated with it. Devorah mentioned, for instance, that when her oldest daughter gave birth to a son, she took the baby in her arms but did not feel a strong attachment to him because it felt like just another baby. She quickly added that she was certain she would love the child in the future but didn't see herself as one of those grandparents who couldn't wait to have grandchildren. Her view expressed the mix of emotional attachment and instrumentality that shapes the path of prolific reproduction. As her statement also reveals, having large numbers of children extended by many years the period of a woman's life devoted to caretaking. Often older women cared for grandchildren, sidelining other pursuits or interests. In some cases, a woman's youngest child was so close in age to her first grandchild that she would be caring for both generations at the same time.

A recent film depicting the constraints and conflicts of religious motherhood confirms many of these ethnographic observations (Lori 2009). *Shira* portrays the tribulations of being a devout Jewish woman in a patriarchal Jewish context. The film is modeled on the personal experience of Miryam Adler, a Russian filmmaker from Kiryat Arba, who married at age eighteen and had become a mother of six by the age of twenty-eight. Adler took the unusual step of attending a film school in Jerusalem when her youngest child was just over a year old in order to produce a film showcasing a religious (settler) point of view.[19] In her film, like the earlier Leavitt diary, the wider context of settling for ideological reasons recedes into the background, and the dilemmas of becoming a mother in a religious community are foregrounded.

Shira depicts the plight of a devout Jewish woman and mother of five young girls who does not want to have another baby right away. Her husband reminds her that she has not fulfilled the mitzvah (religious requirement) of having given birth to a boy, and he urges her to keep trying to become pregnant. The pressure on Shira to continue having more children comes from a patriarchal religious context where women are often subject to the wishes of their husbands and rabbis. It also comes from the lack of communication between husband and wife in an arranged marriage. In the face of exhaustion and depression, Shira falls asleep in the middle of the day while her children are left unsupervised, leading to a deeply ominous scene that almost ends in a death or injury. Later on in a playground surrounded by more children, Shira's pregnant friend advises her to take control of her fertility by lying to her husband about the duration of her menstrual cycle and by prolonging religiously required periods of sexual abstinence in order to prevent getting pregnant.

The film is loosely autobiographical. Adler mentions that she made the film to reveal that not every woman is capable of being the mother that the religious community envisions she should be and in order to prevent other women from suffering as she once did. As a religious filmmaker, however, she nevertheless seems to uphold many of the traditional roles she has been called on to fulfill. For example, Adler doesn't entirely set aside her role as a mother but rather uses three of her own daughters as actors in the film. The boundary between public and private vanishes in much the same way that it does in maternalist protests, using mothers in caretaking roles to advocate for the takeover of Palestinian land.

From National Demographics to Religious Reproduction

As much of the scholarship on Israeli pronatalism makes clear, demographic considerations appear early on in both Zionist and Israeli nation-building practices (see, in particular, Kanaaneh 2002; Portugese 1998). While a variety of cultural and historical reasons account for this, the need for a Jewish majority has been linked to an array of rights excluding or marginalizing Palestinians in the context of a Jewish state. Though most nationalisms depend on creating distinct populations, Zionism's particular emphasis on demographics emerged because of the contradictory way a Jewish majority was constituted: namely, from diasporic populations immigrating to a territory inhabited by a people mostly forced to leave upon the establishment of the state. Thus, at a structural level, pronatalism and immigration formed the cornerstone of the demographic focus within the Zionist project.

In the Yishuv period prior to the founding of the Israeli state, demographics and security were intertwined. Gershon Shafir (1989:187–98), for instance, argued that the demographic emphasis (and ethnic closure entailed in it) was critical in forming stable communities through a "conquest of labor" strategy used by the Jewish immigrant agricultural workers who, unlike others at the time, remained in Palestine and defended their communal agricultural settlements against security breeches and the use of far less expensive Arab labor. Many of Israel's founders (most prominently Ben Gurion) subsequently emphasized women's obligations to bear sons for the nation and left a legacy of state-sponsored pronatalism, including monetary incentives for producing large families, the relatively late legalization of abortion, a health-care system with extensive fertility coverage (barring contraception), and the largest number of fertility clinics per capita in the world (Kanaaneh 2002:28–43).

Ideological settler communities have appropriated this demographic emphasis and combined it with religious commitments to encourage women to participate in creating a Jewish settler majority population in the Occupied Territories. Ideological settlements, then, should be construed not only as strategically arranged housing blocks and territorial strongholds but also as devout communities invested in expanding a settler presence through reproduction while at the same time creating deep-seated attachments to a locality. As with incursions into the Hebron mosque, the founding of the Jewish cemetery, the takeover of the Daboya building, and the cultural representations of religious motherhood that circulate, gender is an important ground of the valued and exclusive social bonds projected onto the realm of territory.[20] It is not the Bible alone that gives rise to resolute attachments to sites of Jewish origin. As the cases of Kiryat Arba and Hebron explicitly demonstrate, the aggrandizement of maternal roles creates opportunities to extend emotional bonds from children to newly claimed strongholds and outposts, where attachments between people and place are expediently forged in Palestinian areas that have little remaining material evidence of a Jewish past.

Chapter 4

Spaces of the Everyday

It is no accident that every Sabbath, the Palestinian residential area that lies along the road adjoining Kiryat Arba and Hebron's most contested religious site, Meʿarat ha-Makhpelah/al-Ḥaram al-Ibrahimi, is descended on by settlers walking down from Kiryat Arba. Moreover, thousands of devout settlers throughout the West Bank convene on the area during the Sabbath when the Torah portion that refers to Kiryat Arba, *Ḥaye Śarah* (Life of Sarah), is read aloud. The same happens during Israeli election campaigns and Independence Day, when settler groups are bused in on chartered buses. On these days, Kiryat Arba, the often-maligned marginal settlement, revels in its glory as a vanguard community that has renewed and secured the Jewish site of Hebron. These ongoing forays of armed ideological settlers into Palestinian space are in part a spectacle of power, inviting comparison with a military formation, which displays its might through its very arrangement and uniform movement.[1] Yet among groups of religious settlers converging on Hebron, the self-presentation is quite different. Here the community, or familial crowd, moves quickly through space toward a sacred religious center. It is a familial crowd in the sense that women and their many children outnumber adult men. Women push strollers at the center, while their kids and husbands surround them. The weaponry enclosing these women and children is not always plainly visible as in a military formation but rather appears randomly; a handgun in a holster, a bulge under a shirt, a rifle, or an Uzi submachine gun casually slung over a shoulder. Arms appear to be woven directly into the fabric of the religious community. In addition, Israeli soldiers dressed in uniform are positioned at strategic points on the road, mainly at blind corners and on rooftops, to insure the safety of this moving crowd. The tone of these incursions is a strange hybrid of militancy and festive religiosity. Families decked out in their finest, women in colored hats and flowing skirts stride next to their husbands in simple white shirts and solid pants, there to see and be seen. But they proceed as if they alone are visible to one another.

While walking resolutely and socializing, on the few occasions I observed, people avoided making reference to the Palestinian residents they were passing, to the layout of physical space around them, or to any details of the observed landscape. Sometimes a donkey would stray into the road, chased by its Palestinian owner, and the armed settler crowd would momentarily move around both, continuing to walk without acknowledging the incident. Indifference was the disposition that allowed settlers to navigate directly through the winding roads of Palestinian areas. With this reduced display of affect, space would be routinely overtaken, a space not literally devoid of its inhabitants, but one in which Palestinians were completely disattended and pushed into the background as if an element of the landscape. Coupled with arms, which the state readily distributes to male settlers who live in the West Bank, are a series of curfews applied to Palestinians living in the area. In Hebron, these curfews confine Palestinians to their homes for long hours, while allowing settlers to move about freely. It is often the case, then, that while a settler crowd is moving through and even festively overtaking Palestinian residential space in order to celebrate a religious or national holiday, Palestinians are made into unwilling spectators of this display of domination.

Making the self visible in the form of a crowd marks a transition between walking as an everyday use of space and the collective control of space as a moment of ethnic assertion. While other observant Jewish communities routinely walk to the synagogue on the Sabbath because they are forbidden from driving, here walking through Palestinian neighborhoods with arms in view of an occupying military force confers on Sabbath the added sense of a permissible incursion. A similar settler presence during a long line of Jewish holidays, national celebrations, and political rallies serves to permanently alter the character of Palestinian space, turning areas where settlers congregate into Jewish ethnic enclaves. Jewish settler routes that connect formal and informal enclaves, as well as Kiryat Arba and other nearby settlements or illegal outposts in the area, are often incrementally expanded through movement or storming practices and retroactively secured by a military presence. These secured routes in turn create opportunities for other settler incursions, expansions, and appropriations. Ethnicized space, in other words, does not adhere to permanently fixed boundaries or legal limits. Rather, it is always in the process of being fragmented and reassembled in micro-level activities and

Figure 9. Settlers walking through Hebron on the Sabbath in front of Palestinian youth watching in the background, 1996. Photo by author.

Figure 10. Likud election rally in front of the Tomb of the Patriarchs, 1996. The words on the back of the T-shirt mean either "I'm certain of Netanyahu" or "With Netanyahu, I'm secure." Photo by author.

clashes, underwriting settler solidarities, on the one hand, and Palestinian otherness, on the other.

This chapter focuses on the ways in which ethnicity is deployed and transformed in ideological settler uses of space, as well as how it aligns with the textual, symbolic, and interpretive reorientations of Jewish tradition taking place. As the classic work of Fredrik Barth (1998) has shown, ethnicity is not, as often imagined, simply a set of received cultural traits handed down and inherited from one's ancestry. Rather, it is part of an active process of differentiation being shaped through social interactions with the "other," using cultural markers and meaningful practices to organize "difference" into collective solidarities (ibid.). Moreover, in ethnically marked or segregated spaces, these expressions of difference, which take the form of religious practices or those connected by blood ties, language, and territory, are often read directly through spatial arrangements that convey social asymmetries (cf. Prentiss 2003; Peled 2014).

Barth's work on ethnicity does not address overt displays of militant ethnicity in parades or incursions, yet his focus on processes of engagement with the other as an organizing principle of ethnicity is suggestive for understanding how the solidarities forged among Jewish settlers are not just a given aspect of devotion but instead are actively shaped through antagonistic encounters. His work also suggests ways in which space itself can be marked by these encounters. In other words, longing not for the messianic era but for direct control over the space of the other, as well expansion of Jewish settler spaces already inhabited, becomes the basis for a distinct form of collective solidarity that differs from other forms of Jewish ethnic expression.[2] These processes of forming ethnic enclaves, then, underscore an evolving fundamentalist Jewish sensibility, because, to draw on the insights of the geographer Roger Stump (2000), they insulate the religious community from contacts with outsiders and provide it with a sense of purity.

Independence Day and Its Relation to Hebron

In order to look more carefully at the nexus of ethnicity and religious right radicalism emerging on the occupied periphery, I examine ideological settler celebrations of Independence Day over the recognized boundary of Israel, paying attention to the ways settler identifications tend to overtake standard expressions of nationalism on this holiday. The commemoration of Independence

Day in Hebron has a different valence from similar celebrations taking place within the boundaries of Israel proper.[3] It is distinctive not only because religious settlers have a divided allegiance to the nation but also because it highlights local settler attachments to an embattled religious site. In the culmination of one observed Independence Day celebration in Hebron, which took place at nightfall in the open square facing Me'arat ha-Makhpelah/al-Ḥaram al-Ibrahimi, the partitioned mosque was symbolically overtaken by a demonstration of a distinctly ideological settler form of belonging. Lit by a dramatic spotlight, an oversize Israeli flag was ceremoniously raised using a rope pulley mounted on the site's massive Herodian-era wall until it fluttered over its dome and beneath two minarets. Once flying overhead, the flag was offset by a sophisticated display of fireworks, financed by the municipal council of Kiryat Arba and reportedly among its greatest expenditures each year. During the festivities, the flag and fireworks transformed the mosque into an atmospheric backdrop, rather than presenting it as a site shared with Muslim worshippers. Using the Israeli flag, then, allowed these settlers to participate in national celebrations taking place around the country and, at the same time, claim the entirety of this particular site as a vehicle of shared affiliation.

A similar form of erasure took place earlier that day, when a crowd of several thousand devout settlers from other settlements converged on the area and stormed through Palestinian spaces with a sea of blue and white Israeli flags. Palestinian residents living in this section of Hebron, meanwhile, remained under curfew, observing through their windows and doors. In the midst of this one-sided effervescence, an Israeli soldier stationed on the roof of a Palestinian house looked down and noticed some of his devout friends, a group of approximately seven or eight yeshiva students wearing knitted *kippot*.[4] They began conversing, shouting back and forth between the road and the roof, until the soldier gestured for them all to come up and join him. As the Palestinian owner of the house and his children looked on, these religious students started to file into his front yard and climb up a provisionally placed metal staircase put there for military use. An older Israeli policeman, stationed on the ground, whose authority and arena of control had been ignored, lost his temper: "Do you think this is your *own* property? This is an entrance to a *private* house," he shouted at the soldier and settlers, upholding the principle of private property as inviolable.[5] He then went on to chastise them further: "If there were stones down here, I'd throw them at you myself," ironically claiming solidarity with Palestinian acts of stone throwing during the First Intifada, or Palestinian uprising.

While the policeman's demand that the religious students leave the area was intended to impose a semblance of law and order, these limits seemed entirely arbitrary in the context of the occupation where any soldier could decide to set up a lookout on any roof of a randomly chosen Palestinian house. The policeman was there, however, to set some limits by allowing settlers to move through the area but not in ways that might spiral out of control. As the religious students were opening the gate of the Palestinian owner's yard, they gave no indication of being overtly hostile. Rather they simply proceeded as if they were engrossed in a lively conversation with their friend, whom they happened to bump into in the crowd. From their vantage, entering a yard seemed no different from traversing other Palestinian areas because crowds of religious settlers were already moving through the streets with the protection of soldiers and the sanction of the police. While a massive settler presence converging on the area fell within the parameters of what was allowed, the actions of these religious youth did not. Moreover, the soldier's presence on the roof was not deemed provocative by police authorities, while calling up others to join him was off limits, an illustration of the how the gray zone operates in practice. Similarly, while the fireworks celebration reveals settler uses of national symbols for controversial claims, this moment highlights the ways settler incursions into Palestinian areas can appear as routine practices (and entitlements) that settlers engage in when inhabiting the area. This is not to minimize the many documented cases of settler harassment, direct violence, and destruction but rather to point out that trespass by settlers in this context has become almost automatic and engrained because of the spatial inequalities that have already been established.

In addition to assertions of solidarity, Independence Day on the periphery also attempted to shift the ethnonational underpinning of the Jewish state into an overtly religious register. While inside Israel, the celebration was replete with patriotic and declarative overtones, in Hebron it became an occasion to leverage trenchant critiques against the Israeli state while appropriating the authority of Israel's national symbols in order to do so. On this day, for instance, radical rabbis proffered the message that Israeli independence had not actually been achieved and that settlers were calling for its completion through the religious requirement of having all the people of Israel return to their entire biblical homeland. In this way, ideological settler "independence" participated in aspects of Israeli nationalism, but ultimately endowed it with a distinct religious cast that was ultimately at odds with more standard versions of it. The speeches by rabbis and municipal council members had a local inflection that was not easily integrated into any wider preoccupations of the nation, focusing on the specific accomplishments and hardships of a distinct ideological settler sector. Prayer at the Tomb of the Patriarchs took place during Independence Day, further shifting the national register to a more religious tenor. The service included religious lyrics sung to the tune of the national anthem, while in Kiryat Arba itself, military tanks were stationed inside the settlement so that devout kids could play on them. Most families in the settlement also participated in a Sabbath-like meal at home. At one such celebration, wine was poured, and challah bread was broken as if to mark the beginning of a Jewish holiday, even though it was really not. Therefore, the bread wasn't distributed by a senior male member of the household standing at the head of the table, and the electricity wasn't turned off, so the radio blared patriotic music during the meal. For this reason, the day seemed like an oddly disjointed form of Jewish observance.[6]

The Spatial Character of Ethnicity

Ethnic exclusivism in Israel has often been linked to Oren Yiftachel's (2006) use of the term "ethnocracy," referring to the state's project of expanding into and taking control over territory, while overlooking its (legal) obligations to a Palestinian citizenry. In his use of the term, collective solidarities and a sense of ethnic superiority are mutually reinforced by the state's expanding control of land. Yiftachel contends that the Judaization of land is mainly a state-driven process, and it paves the way for religious radical movements on the periphery. Moreover, he contends that Israel's territorial expansion can be traced to the hegemony of an Ashkenazi elite, which advances its own "'ethnicizing' political and territorial agendas over contested space and power structures" (ibid.; Yiftachel and Roded 2011:179). Israel's ideological settlers are, in this view, pawns of these (elite) interests, acting to merely implement the state's "ethnocratic" agendas rather than remaking them in any significant way. Yiftachel's structural and state-centric descriptions, while valuable for illuminating conditions and resources that impact a settler sensibility, nevertheless tend to obscure the range of practices and ideological points of view that motivate religious settlers to act as they do on the ground. They also minimize settler agency and their directly antagonistic ways of driving the ethnic dimension of the state toward even greater exclusivity and violence. As these ethnographic illustrations show, religious settlers are not simply implementing an existing structure of exclusion but changing its very ground of legitimation by adding to it biblical imperatives that are linked to sacred sites and collective solidarities produced in direct encounters. Through these means religious settlers actively replace established ethnonationalist loyalties and forms of exclusion with those that have a distinctly militant religious modality.[7]

Many other investigations of ideological settlement remain at an abstract conceptual level in that they take, in one fashion or another, the relationship between Jewish belief and a settler's felt attachment to land as a given or starting point rather than as a true object of inquiry in itself (cf. Ravitzky 1996; Feige 2009; Sprinzak 1999). They move (analytically) from resolute attachments to an analysis of Jewish concepts (e.g., inheritance, possession, ingathering) and messianic longing linked to the Land of Israel, bypassing many of the social (colonial) practices and power asymmetries that help sustain these metaphysical convictions in an occupation. The absence of a contextualized and practical engagement with the ways a religious belief system is actually being supported by a distinct social context results in the figure of the "fanatic" as an explanation in itself and a preoccupation with psychological flaws characterized by excessive certainty alone, leaving the impression that irrationality is the main culprit.

Robert Paine (1995), in contrast, sets out to map the contours of a devout settler's "certainty" in a different way. First, he notes that the relationship between "certainty" and "place" is co-constitutive, and that certainty brings with it a set of distinct moral commitments (e.g., the land can never be returned). He then highlights the many ambivalent and contradictory ideological fields embodied by Judaism, Zionism, and Israeliness, revealing how these categories overlap, blend, and depart from one another with respect to place and territory. For him, Israeli and Zionist certainties are not only produced in the face of political *uncertainty*, they derive from attempts to implement a utopian project, giving "territoriality to a de-territorialized people" (Paine 1995:180). His analysis is mainly concerned with intra-Jewish/Zionist discussions, and he points out that in defining a distinct place of their own, religious settlers focus not so much on enemies of the Jews as on Jewish enemies, that is, secular Israelis who see Judaism as heritage alone. For Paine, the religious certainty of fundamentalist settlers is a derivative of the certainty of the Zionist project. He sees settlers, then, as shifting Zionism's resolute character onto a "metaphysical plane," giving its certainties a religious orientation and more authoritative form (175).

While Paine's essay alerts us to the significance of the spatial realm in relating "ideology" to "place," it does not move beyond discourse, leaving aside the issue of how space shapes the realm of religious practice and elicits violence in a militarized and colonial context. For this reason, I consider the ways the certainty of religious settlers in claiming sanctified Jewish places links directly with the sum of their "ethnicized" spatial practices in occupied Palestinian areas. In doing so, I aim to explore the conditions and sensibilities that spur religious radicals to use violence, seeing its ethnic dimension not as given or exceptional but rather as a core element that needs to be explained.[8]

The remarks of Eli, a nine-year-old boy, standing on the Jewish side of the colossal stairway of Hebron's divided seventh-century mosque illustrate how ethnic exclusivity fuels a distinct form of religious certainty in this context. Eli was deeply devout and bordering on anemic from extensive Torah study. As residents of the tightly knit religious settlement of Kiryat Arba, mostly men, passed in or out of the Tomb of the Patriarchs during noon prayers, they slowed to greet him. Striking up a conversation, I remarked, "You sure know a lot of people," to which he screwed up his face and insolently replied, "Of course, I've lived here for nine years!" When I responded that nine years was indeed a long time, he countered: "It's not so long—I could stay another twenty!" As we were talking, a group of approximately forty Palestinian elementary school boys, who seemed close to his age, walked down an outer road chaperoned by their teachers. His attention locked onto them. "They have such big classes!" he remarked. "I wonder what they're doing!" Without pause, he continued: "They are going to have to leave here." The assuredness of his statements exceeded his age. I responded, "Where would they go?" and he answered matter-of-factly, "Syria, Lebanon, Jordan." Then he launched into a long soliloquy in order to inform me, an adult neophyte, about a few basic givens. Pointing to the crumbling structures that surround the Tomb of the Patriarchs, now a no-man's military landscape of ruined buildings that Palestinians have had to leave, Eli switched to Hebrew: "Look at this place—it's turning into a garbage dump [*paḥ zevel*]. In about twenty years, the Arabs are going to run out of money [*ha-kesef yigamer*] and the whole place will turn into a big dump. See those houses there? Eventually we're going to rebuild [*le-shakem*] them. Ariel Sharon was here, and he promised that he would connect up Kiryat Arba with the Jewish Quarter in Hebron!"

It was unusual to hear such a blanket statement of intent, but as he spoke, I was reminded of much that remains tacit in ideological settler claims to sacred places—along with a devotion to these biblical sites and Jewish origins, religious settlement is predicated on assertions of entitlement in the face of difference. The religious sensibility of hard-line settlers is colonial in the sense that it provides a religious basis for the ideological position that Palestinians must depart (aside, they say, from those who agree to live under Jewish rule), making their property available for "renewal." Yet at the same time, the relative privilege of these settlers depends on the very presence of a Palestinian population, posing an existential dilemma for them that can never be resolved. Coming out of the mouth of an imperious nine-year-old, the assertion that Palestinians would have to leave struck me as both a bit sad and childish for the way it parroted more sophisticated versions of ethnic exclusivism. Eli was a staunch believer in these matters but, as a product of his community, not entirely by choice. For this reason, his grand pronouncements elicited in me both aversion and a strange pathos: *The Arabs would all leave. Hebron was moving forward as planned and even supported by a politician as prominent as Ariel Sharon himself.*[9] I quietly listened and made a mental note of it, though I was torn over whether to challenge him more directly. I refrained, believing at the time that my question had been enough.

One of the central ways the settlers of Kiryat Arba participate in the practical process of increasing their control, demonstrating their belief in the occupancy of the land, is through everyday uses of space. Their spatial practices, short-lived moments in a larger spatial logic, have the significance that they do because they invoke, implicate, and enhance the power of the state, but at the same time they shift an older nationalist problematic onto a religious register that often redirects it. If settled religious spaces are also ethnically exclusive enclaves, the practices through which they are made need more careful elaboration. I intend, therefore, to focus on ethnographic examples of settler uses of space and antagonistic small-scale incursions that fall under the radar of most authorities.

Settler Presence and the Repertoires of Indifference

The spatial character of ethnic exclusivism is clearly illustrated in the limited but routine interactions that take place between ideological settlers and Palestinians on the boundaries of Kiryat Arba, though these interactions often remain unacknowledged. This is primarily because the settler stance is one of nonrecognition and indifference to the presence of Palestinians living adjacent to them. Emotional indifference toward the other, the limits of affective engagement, is not uniquely characteristic of settlers but is nevertheless a striking feature of this context, particularly given the deep emotional ties that settlers have developed with many sanctified places.[10] Though Eli's comments did focus on the Palestinian presence, many other Kiryat Arba settlers tend to go to considerable lengths to avoid any mention of this "adversarial" community, except during outbreaks of violence. This form of nonrecognition and the stark social divide it entails, however, is a fragile fiction. As an example of this fragility, consider the stretch of cyclone fencing topped with circular barbed wire that surrounds much of Kiryat Arba. Though Palestinian homes and families are visible on the other side, and the sidewalk of the settlement runs alongside of it, it is common practice for residents walking within the settlement to ignore Palestinian residents living across the fence. On occasion, however, someone would note that a Palestinian house had

Figure 11. View of Hebron mosque and Palestinian Hebron from behind a section of the security fence in Kiryat Arba, 2011. Photo by author.

been expanded or that the vicinity seemed to be more populated than it once had been.

One Saturday afternoon in Kiryat Arba, I spoke with an informant and his three children as they were returning from praying at Meʿarat ha-Makhpelah. For kids as well as adults, walking through Palestinian areas to pray at Meʿarat ha-Makhpelah takes on the character of a celebratory outing permitted during the Sabbath. In keeping with the day's festive spirit, they invited me to join them in visiting the chicken coop, which I had heard about but not yet seen. The coop was simply a small fenced-in area that to my mind seemed to replicate in miniature the larger enclosed space of Kiryat Arba itself. The builder of the coop, however, intended to instill a broader sense of nature and to provide an experience that would generate what he called an "environmental awareness" among the settlement's children, which he felt was greatly lacking.[11]

We followed a small path through a thicket of bushes until there appeared a low but unmistakable coop filled with a few isolated birds. The kids stood peering into the mesh wiring at the chickens until a rooster suddenly inflated his chest and produced the loud triumphant sound of crowing. Then from the Palestinian side, in sequence as if in conversation, another rooster crowed in response. This exchange created a slightly awkward moment because nobody seemed to want to acknowledge the exchange. Then once again the Kiryat Arba rooster crowed, and the rooster on the Palestinian side crowed in response. This time, the kids doubled up in laughter, now dispelled of the idea that the birds would follow the stark divisions they as settlers took for granted. The father, on the other hand, sardonically remarked to me that the situation reminded him of Majdal Shams.

Majdal Shams is a Syrian (Druze) village located not in the West Bank but, rather, in the Golan Heights along the Israeli-Syrian border. Following the 1967 war, the Syrian village was arbitrarily divided, and a tall cyclone fence was installed overnight, splitting families, friends, and relatives between Syrian and Israeli rule. Living (as a divided town) for years without telephone connections or mail service in two countries without diplomatic relations, families poignantly would meet (before Internet access) at the double fencing of the border. New parents raised infants over their heads for the other half to see, and families shouted through booming megaphones to communicate news to one another. The arbitrary border forcing the indefinite separation of family members was one of the many tragic consequences of random cease-fire lines made into final borders. For this Kiryat Arba informant, however, only the Druze of Majdal Shams, rather than the Palestinians living adjacent to him, were the casualties of arbitrary borders, captive to separate and hostile nations. Palestinians across the fence were his direct adversaries and not considered to be living with similar divisions or as refugees.

This incident illustrates the degree to which nonrecognition and indifference create routine erasures within the colonial arrangement that shapes a religious settler context. It has antecedents in the "frontierism" of pre-state settlement, where "the frontier" refers to liminal zones of collective control.[12] Baruch Kimmerling (1983), using an inverted Turnerian thesis, argued that among those in the Yishuv (1882–1917), the *unavailability* of free territory, or "low frontierity," in Palestine gave rise to strong collective bonds among Jews. Their mode of settlement was initially shaped by the structure of Ottoman land ownership, he asserted, since tenancy and cultivation became the basis for the de facto ownership of *miri* lands (i.e., lands leased in usufruct).[13] The equation between being present and owning, the ability to requisition absentee Palestinian land and render it state property, as well as the consolidation of vast tracts of state land were remade into key features of a nation-building process. Yet in contrast with this period of Zionist settlement, the solidarities of ideological settlers have been built on unconstrained expansion, and, in view of the occupation's legal gray zone, they have also been forged by legal privilege, ongoing antagonistic engagements, and direct violence against Palestinian residents. The ethnic underpinning of settler solidarity in Hebron entails not only a quest for the control of more land but effacing the relational qualities and interchanges that regularly constitute basic notions of sociality.

Asymmetrical Movement and Segregated Space

In view of these forms of indifference, I was surprised to learn that a small number of settlers from Kiryat Arba actually bought goods in Palestinian shops located across the road from one entrance that was secured against nonresidents by an electronic gate and guard booth. Rivka, a flamboyant Parisian in her early seventies, who often recounted her experiences fleeing Vichy France, for example, used to step outside of the settlement to shop at a Palestinian dry-goods store. She was considered to be a marginal member of Kiryat Arba by virtue of her eccentric solitude, dire poverty, many pets, and mystical tendencies in religious matters. Yet because she had lived in the settlement for more than sixteen years, she provided key insights into its contradictions and ironies, emphasizing among other things current waves of gossip and scandal. Even though many other settlers disapproved, she purchased her produce at the Palestinian shop beyond the gate because its vegetables were much cheaper and of better quality than those available at the supermarket in Kiryat Arba, which had been trucked in from Israel. It was not that she had any lesser conviction about the rights of Jewish settlers to control all of Hebron; rather, when it came to shopping for vegetables, she took a pragmatic approach.

If my errors might reflect badly on her, Rivka would adopt a didactic tone and tip me off on what was deemed appropriate. Once when I accompanied her to this shop, she stopped before the entrance and pointedly remarked, "Here, you don't say hello." When I responded, "Why not?" she simply insisted, "You just don't," and changed the subject. Upon entering the store, I noticed that there were several other settlers from Kiryat Arba buying items and that it was not as unusual to shop there as I had imagined. Yet while settlers and Palestinians momentarily inhabited the same space, they adhered to a code of highly guarded behavior. For settlers, it was as if the fenced-in boundary of the settlement, with its proscription against recognition of the "other," was extended beyond its original location. Within the shop, settlers and Palestinians did not greet one another, and they deliberately did not make eye contact. So even as the settlers of Kiryat Arba spoke as needed with the Palestinian shopkeeper and his employees and were acutely aware of their presence, they maintained their stance of nonrecognition. Their modes of indifference were therefore transposed to the arena of small transactional exchanges.

Even in these economic transactions, settler claims to belonging were being asserted. As Rivka rifled through the crates of tomatoes, cucumbers, peppers, and grapes, she asked the shopkeeper in accented Hebrew, his back turned to her while arranging the shelves, whether the vegetables were *baladi*, appropriating one of the Arabic words she knew. Hebrew itself was not a language she had mastered, but she seized on the term "rural" or "native" (which derives from "country") to inquire if the tomatoes and cucumbers had been grown locally. She had uttered *baladi* partly as a matter of emphasis, partly in order to make herself more readily understood to the shopkeeper in his language, marking their differences. Yet this question raised the issue of what she meant by "homegrown," particularly given that whose home it was and who was "native" to the area were matters of grave dispute. I thought I detected a hint of irony in the Palestinian shopkeeper's affirmative response, but this may have been my own imagination at play. While Rivka's engagements could be seen as participating in an earlier Zionist ambivalence that treated Palestinians as both absent and romantically rooted in the land, her warning to me not to greet the shopkeeper was a more pointed lesson in actively erasing his presence. I therefore see her interaction as double-sided: a way of diminishing her outsider status through practical deeds that tied her to a place of origin and as an incursion into another's space.

The Legibility of "Difference"

Though this example illustrates nonrecognition in the case of a purchase, I complicate the way "otherness" is produced by considering another transaction, this time focusing on the ways segregated space further entrenches ethnic divides. Shimon, the owner of a Kiryat Arba studio I had rented for a period of research, shopped directly in the Old Palestinian market in Hebron before the expanded presence of settlers in the Jewish Quarter nearly closed it down. His case is instructive because his Yemeni background meant that as a settler he occupied an in-between position as both colonizer and colonized. As a traditional rather than observant Jew, he maintained that he often frequented the Palestinian market by passing under the radar of the stark social divisions that were operating. He still benefited from the legal rights and economic privileges afforded to other settlers at the expense of Palestinians, but he expressed a certain affinity (mainly using irony) with Palestinian and Arab culture beyond the borders of the settlement.

Shimon's family had come to Palestine from Yemen before the establishment of the Israeli state, and he mentioned that he had grown up in a mixed Jewish and Palestinian neighborhood in Jerusalem before moving to Kiryat Arba. Fluent in Hebrew and Arabic, he would often go into the Palestinian marketplace to purchase food and other household items, claiming that it was more convenient than taking the bus all the way into Jerusalem. When I first arrived to carry out my research, I needed to purchase a small gas burner and desk, and it was Shimon who offered to help me get these items, insisting they could be easily and cheaply acquired. I assumed, given the steady influx of Russian immigrants then arriving in Kiryat Arba, that we were going to buy these things in a used appliance or furniture store within the settlement, but instead I ended up accompanying him on one of his expeditions into the Palestinian market.[14]

I met Shimon at a bus stop and instead of getting off in another section of the settlement as I had intended, he asked me to wait until it stopped at the heavily guarded Jewish enclave in Hebron. In some sense, there is no better place to begin a close examination of settler movement and (in)difference than by considering one of its most important vehicles, the bullet proof bus route connecting Jerusalem to Kiryat Arba and Kiryat Arba to the Jewish settler enclave directly within the Old City of Hebron. Separate Palestinian and Israeli bus services are used to transport people to either Palestinian villages or Jewish settlements in spite of the fact that in most cases, these locales are situated in walking distance of each another. Since the majority of people living in Kiryat Arba do not own cars, Bus 160 is the lifeline that links not only different fragmented neighborhoods within the settlement to one another but the entire settlement of Kiryat Arba and its neighborhoods to outlying settlements overlooking other Palestinian agricultural lands and these areas in turn to the city of Jerusalem. Part civilian transit, part metallic military vehicle, the windows of the bus are fortified with thick shatterproof panes of glass, as well as by a heavy steel grate that is affixed to the front of the windshield.[15]

From inside the bus, a settler's visual perception is reduced to two possibilities: one that peripherally takes in the moving landscape of Palestinian villages and variegated fields with the dreamy remove of the distortion caused by filmy glass; or, alternatively, looking directly through the windshield into that cage-like grid and seeming to be directly under siege. En route to Kiryat Arba from Jerusalem, the bus passes through a permanent military checkpoint marking the boundary of the West Bank and then stops at a number of Jewish settlements until it halts midway for a military vehicle to come and escort it to the entrance of Kiryat Arba. The distinct experience of sitting in Bus 160 is one of being waved through a series of checkpoints—the very junctures where long lines of Palestinian vehicles, their occupants being questioned one by one, wait and move through one at a time, sometimes standing for hours. Movement itself and the complex relationship between agency and moral favor that results from it become a tangible marker of the divide between Jewish settlers and Palestinians under occupation.

It was this absolute difference that Shimon sought to subvert and, by doing so, to stake out a distinctive location for himself, relying on Arabic as a bridge beyond a mainly Ashkenazi (European Jewish) form of settler colonization. In doing so, he disregarded his community's prevailing stances of indifference and opted for more direct interactions with Palestinians. Chatting with amusement and playfulness on the final leg of the bus ride into Hebron, Shimon remarked to me that dressed as he was, gesturing dismissively at his plain clothes, no one would be able to tell whether he was Arab or Jew. Then eyeing the pants I was wearing, he added that I would blend in as a tourist. I wondered what it was about his dress or my own that led him to make these confident assertions. He was dressed no differently than at any other time I had seen him—modestly and in clothes that were clearly not new but not shabby either. Perhaps he was referring to the fact that he did not carry arms, nor was he wearing a knitted skullcap or any other religious marker. Yet as we walked into the Palestinian market, I began to have serious doubts not only as to whether his identity was as ambiguous as he assumed but also as to whether my own would be seen as that of a tourist.

Among settlers in Kiryat Arba, Jewish ethnicity and religious devotion are intertwined, even as both are in the process of being co-constructed. Even if Shimon was not readily identifiable as a settler because of his Middle Eastern background and relative secularism, merely seeming to not belong was enough to draw the attention of Israeli soldiers deployed in the area. He first stopped at a small Palestinian fabric shop where he effusively greeted the owners and then arranged to have one of their employees drive us back to the settlement. I am not sure in what capacity Shimon had made their acquaintance, and whether or not they were reluctant to comply with his wishes, but they sent their employee out with an old car to deliver the goods that had been purchased. No sooner had Shimon put some of the purchased items into the trunk than two edgy soldiers treading through the market stopped us. Peering out from under their helmets and dressed in full combat gear, the soldiers immediately asked him whether he was Jewish. Shimon responded curtly that he was and glanced away, apparently not wanting to make overtures of friendship to the soldiers in the context of the crowded Palestinian market. He seemed irritated as well, perhaps because he had been singled out in a public way—directly in view of a group of shoppers who were carefully observing the interaction.

One of the soldiers then asked whether he wasn't afraid to be in the market, and Shimon replied assertively that it was no problem at all. Exchanging incredulous looks, the two trudged on. By his response alone, Shimon had placed himself squarely within the settler camp. Had he indicated that he was not Jewish, or that he was indeed afraid of being there, the soldiers might have asked for identification and detained him. Not acknowledging fear, though, was squarely in keeping with the settler ethos of being "at home" in the area. In this respect, other settlers often mentioned that seeming fearful would be a concession to the idea that they didn't actually belong in Hebron, and some even looked down on those who walked around with a gun.[16] My interest here, however, is not only in the affective dimensions of a settler sense of belonging but in the way trajectories of movement themselves can be officially read as markers of an ethnically based identity and the manner in which these in turn eventually create spaces that are deemed permanently Jewish. In stopping Shimon, the soldiers curiously did not ask him for any proof of his identity. Rather, for them, his Jewishness was apparently confirmed by the ways he was negotiating an ethnically divided context. His speech, dress, and mannerisms probably factored into the soldiers' decision not to ask for identification, but, in addition, his identity as a settler was confirmed by his sense of entitlement to walk through the Palestinian market. Shimon may have seemed out of place but he was nevertheless well within the parameters of spaces frequented by other Hebron settlers and therefore squarely participating in the process of remaking Hebron as Jewish territory. With no other direct evidence, Jewishness was derived from the way he moved through Hebron and engaged with these soldiers as if structurally allied with a military force rather than subject to it.[17] The presumption of being able to read who one is from one's trajectory of movement points to the ethnic divisions of space and the marking of difference within the occupation. Apart from being legally designated, this social difference is also constructed through micro-interactions and daily practices, even though it often appears to be part of the primordial difference permeating space and place itself.

Limiting Palestinian Movement

The Palestinian driver's identity presented itself as a problem when moving in the reverse direction. His car bore a blue license plate with the Hebrew letter *ḥet* imprinted on it, making him immediately apprehensible to the guard at the gate of Kiryat Arba as an "Arab" from Hebron.[18] This stark difference was independent of any utterance or gesture that the driver himself had made. The guard, acting in his official capacity, refused to let the Palestinian car into the gates of the settlement. Shimon immediately jumped out and fought heatedly with him, arguing that he had made a prior arrangement that was being ignored. With great reluctance, the guard eventually did let the car in, but not without first withholding the driver's identity card. In the instance of floundering through pockets to relinquish his identification, the colonial aspect of Kiryat Arba could not have been more visible. For this Palestinian driver, giving up his identity card was not merely a formality to get into the settlement but an act of becoming a nonentity within the structure of power distributed over its space. It confined the driver to the settlement, because he would not have been able to leave without an official form of identification. This was a relatively small gesture, and perhaps one that is hardly remarkable within a whole set of antagonistic and colonial exchanges that routinely take place between settlers, soldiers, and Palestinians in Hebron. But it was almost an exact inversion of Shimon's experience, and it caught my attention for revealing the ways space itself played a role in ethnically categorizing social differences. Through identity cards, license plates, military checkpoints, and curfews, Palestinians became tied to shrinking locales while Jewish settlers routinely moved between them and remade boundaries. Moreover, there was no distinction on Israeli license plates between settlers living in the West Bank and citizens living in Israel proper. Sovereign and occupied territories melded together through settler movement and the porous boundary between them.

Michel de Certeau (1984) makes a number of important observations that relate spatial practices to what he calls the legibility of a place, and, in doing so, he offers a way of thinking about the relationship between space as an "enunciative act of movement" and place as a stable effect of these actions.[19] In de Certeau's work, space is made into meaningful places within a broader spatial order that is construed as an ensemble of possibilities. In his Saussurian-like framework, then, the pedestrian (rather than speaker) gives rise to distinct possibilities embedded in the deep structure of space through its routine uses in the course of walking. The pedestrian rearranges existing possibilities and invents others so that the "crossing, drifting away, or improvisation of walking privilege, transform or abandon spatial elements" (de Certeau 1984:98). In the case of Hebron settlers, however, it is important to add to these routine place-making efforts the structural or actual violence of spatial transformations so that spaces become enclaves that are exclusively Jewish. Not only are these Palestinian spaces incrementally being transformed but ethnoreligious "difference" has become permanently arranged within them. Reinforcing settler invocations of a Jewish biblical past, then, are the many ideological certainties that emerge out of an ethnicizing spatial process in deeply asymmetrical conditions and the fervor of belonging that sustains them.[20]

Boundaries and Borders

Ideological settlements create boundaries that become both sites of expansion for settlers and limits of access or mobility for Palestinians. For settlers, boundaries are routinely transportable limits embedded in social practice. For Palestinian residents, on the other hand, the boundary presents itself as a barrier, exclusively safeguarding Jewish settler citizens within a governing structure of economic privilege and legal entitlement. The asymmetry of power inscribed in the process of expanding boundaries is overseen by a military force, making apparent the backing of the Israeli state as a condition for the continued existence of a Jewish settler presence. But while the military ensures settler security, it does not intervene in settler confrontations with Palestinians in these deeply unequal conditions. Direct confrontations between settlers and Palestinians, which frequently occur at the boundaries of a settlement, are significant beyond the locale because they can be invoked or taken up as needed by the state's official plans, projects, and decisions in forging (expanded) national borders.[21]

In the summer months during my research there, Kiryat Arba's boundary areas went up in flames several times a week, prompting a response from the settlement municipality. A fully equipped fire engine immediately appeared on location to extinguish the flames, dousing the area with water until the grassy and stumpy remains were left wet, black, and smoldering. Very often, the embers would rekindle, and only a couple of hours later, the fire engine would reappear. While the cyclone fence was immobile, fire became its counterpoint as an element out of control. It also momentarily blurred the divided areas of "difference" by burning across property and uniting all, Jewish settler and Palestinian, against this natural force. From the opinions, attributions, and accusations and counterclaims that swirled around the fire on the settler side, it was impossible to tell who was to blame for it. Kiryat Arba seemed equally divided among those who thought that the fire was the work of Palestinian adversaries and those who thought that it was possibly the work of its own derelicts. A few settlers thought that dry areas were simply prone to ignite on their own. As the firemen went about hosing down the area on their side of the fence, they did not make great efforts to extinguish the burning area beyond it, leading several Palestinian families to step out onto their balconies and directly face a few belligerent teens investigating the scene from their side.

While the area was still smoldering, the two teens from Kiryat Arba walked up to the fence and instigated a conversation. This was the only direct exchange that I heard during my research, and it coincided with a heated Israeli campaign in the weeks prior to the first victory of Likud's Benjamin Netanyahu over Labor's Shimon Peres. The conversation went roughly like this: One of the teens pointed to a ripped Moledet campaign banner lying in the dirt. With an air of great authority, he ordered a middle-aged Palestinian man standing on his own property to return it, claiming that he had stolen the flag.[22] The Palestinian man, speaking to him in Hebrew, countered that he had not touched it and that what others had done was not his concern. He also said that people living in Kiryat Arba had entered his property countless times. This exchange was prompted by the teen's assertion that something of *his* had been stolen by that particular Palestinian man—if not his own flag, then his by association to Kiryat Arba and the Jewish (settler) community. It was as a proprietor of the exclusionary politics embodied in that flag that he demanded its return, unabashedly asserting his authority. The colonial nature of the encounter was reinforced by the idea that a far-right Israeli party advocating the wholesale "transfer" of Palestinians out of the West Bank needed to be respected by all.

I was taken aback by this brash assertion and the teen's demand, when a second boy, younger and less certain, stepped into the fray. Hearing their taunts, the Palestinian neighbor warned that he had worked in Kiryat Arba for eleven years and knew all of their families well. In response, the younger, chubby and cheeky teen shouted back: "Do you know my family? I'm the son of Baruch Goldstein—ever heard of Yahya Ayyash?"[23] Silence followed. Invoking terrorists, he had disrupted the man's claims of familiarity and civility. Once he had done this, the boy began to wonder and ask, in all earnestness, whom this Palestinian "adversary" would be voting for in the upcoming elections. He did not seem to know that Palestinians under occupation would not be voting in Israeli elections. He probed but did not get any response until he finally stated: "I know . . . you're voting for Peres, right?" Satisfied with his own cleverness, he rocked forward at the force of this allegation. Those who would vote for Peres represented the complete opposite of everything he and other ideological settlers stood for.[24] It was the most extreme order of difference he could imagine, and he mistakenly drew his Palestinian adversary nearer to himself by suggesting that he too was part of an internal Israeli political divide.[25]

The younger teen's error revealed a tendency to categorize Palestinians and Israeli liberals as similar kinds of threat. During the time, all of Kiryat Arba was preoccupied with the possibility of threat as an anticipatory moment of violence. On election day, for example, voting was considered to be an obligation equal to protecting the Jewish (settler) community from imminent danger. A car with a megaphone mounted on its roof drove through the settlement announcing "*pikuaḥ nefesh doḥeh et ha-kol*" (mortal danger requires that you postpone everything), meaning that it was imperative to get out and vote. This was a riff on the halakhic principle "*pikuaḥ nefesh doḥeh Shabat*," which means that the preservation of life overrides other obligations, or, more literally, mortal danger allows the Sabbath to be violated in order to protect oneself and the community. Within Kiryat Arba at the time, voting against Peres was seen as imperative because it would save settlers from the presumed danger of annihilation posed by any territorial compromise.

Reorienting Jewish Law and Rabbinic Commentary

To further explore the manner in which ethnically exclusive spatial practices are believed to be authorized by Jewish law, I turn to a contemporary discussion of halakha, Jewish religious law, by the rabbi (and settler) Yosef Mizrachi. Mizrachi was educated in the *hesder* yeshiva of Kiryat Arba, a long-standing institution of higher religious education that goes back to the settlement's founding.[26] His engagements with halakha reveal a preoccupation with rabbinic debates that presumably authorize a contemporary ideological settler's territorial concerns defined in relation to ethical-legal precepts. Mizrachi's (1974) "Israeli Sovereignty over the Land of Israel in Light of Jewish Law," published as part of a book commemorating the ten-year anniversary of Kiryat Arba (entitled *Kiryat Arba Is Hebron*), begins with a discussion of land as a religious inheritance (*yerushah*).[27] It then explores the rights of non-Jews living under Jewish rule and concludes that any rights designated for the "resident alien" in Jewish law cannot apply to the case of "Arabs." His inquiry is ultimately driven by a preoccupation with finding rabbinic (rather than biblical) justifications for the occupation of Palestinian territory in light of "possessing" and "owning" land, but his line of reasoning actually reveals the many legal precepts in Judaism that are directly at odds with his final conclusions.

The most important divide in halakha (Jewish law) is one that distinguishes between laws that are to be implemented in human time, serving as the basis of an observant community, and those that are to remain hypothetical (including sacrifice, settling, sovereignty) until the messianic era. Maintaining this divide between what is and what is to come, scholars of Judaism argue (Ravitzky 1996; Ricouer 1986), allows for greater reflexivity and social critique. Ideological settlers (and more broadly those in the Modern Orthodox camp), however, understand the founding of the state of Israel to complicate this fundamental divide in religious law. They argue that although the messianic period has not yet been realized, the state of Israel has already fulfilled some of its vital conditions, including the return of Jews from exile to the Land of Israel. This accommodation of the modern state theologically is the impetus for settlers to draw (selectively) from laws on both sides of the divide to suit contemporary ideological aims. Therefore any laws linked to settlement, sovereignty, and possessing the sacred Land of Israel, traditionally deemed hypothetical, have begun to be taken as those meant to be actually implemented in the present. While the idea that the modern state of Israel has impacted contemporary Judaism is not a settler issue alone, ideological settler invocations of Jewish law in light of the state's existence have taken a distinct turn. They have (arguably) pushed beyond Zionism (cf. Taub 2011), which used Judaism to forge a comparatively secular nationalism, and replaced it with an explicitly theocratic political project.[28]

Yet Rambam, or Maimonides, as he is known, presents a major hurdle for settlers. A medieval physician and philosopher of the twelfth century, Maimonides fled persecution in Spain and spent much of his life living in Cairo while treating patients, Muslims and Jews alike, and writing a number of Judaism's central legal and philosophical works. Though Rambam lays out 613 commandments in his canonical work *Sefer ha-Mitzvot* (Book of Commandments) detailing the obligations of a Jewish life, he fails to mention "settling" at all. Rabbinic legal commentators as well as settlers have offered explanations for this omission, one of which is that, for Maimonides, the principle of *yishuv ha-arets* (settling the land) was so fundamental that it was entailed in all the other commandments and therefore didn't need stating. Others, however, have argued that he didn't mention settling because it was not applicable to his understanding of Jewish life, particularly in terms of its cosmopolitan orientation. Still others, including Mizrachi, emphasize the dividing line in Jewish law and discuss the degree to which settling, or living in the Land of Israel (as a theological concept) has any applicability to the historic present.

The main hurdle religious settlers regularly face in making their case for contemporary settlement is that there are multiple passages in these commentaries and the Bible that expressly forbid conquest, or "scaling the walls" of a city and "pushing the end" by forcibly implementing the messianic ideal before its time. Reading the rabbinical opinion of Isaac de Leon, another medieval scholar from Toledo, Spain, and presumed author of *Megilat Ester* who comments on Maimonides, the contemporary Mizrachi writes: "We were commanded, according to what was said at the end of *Ketubot* [Talmud], that we shouldn't rebel against nations in order to conquer the land by force. And this was proven from the verse 'I made you take an oath, daughters of Jerusalem,' which was interpreted as Israel should not scale the walls." Mizrachi then goes on to exegete a series of quotes from these commentaries to discuss the meaning of the term "settle," employing a medieval style of argumentation. Drawing on Aramaic phrases and distinctions such as *peshat* (the surface meaning) and *derash* (allegorical meaning), he concludes that modern Israel has meant that Jews now have a new state of affairs, and that if rabbinic commentators such as Rambam and De Leon knew of it, they too would have supported settling. Mizrachi, in other words, uses the opinions of rabbinic commentators who disagree with him to forge a retroactive consensus of opinion across time in order to legitimate ideological settler aims. In this way, the practical experience of settling in places like Hebron brings new theological understandings and emphases to these rabbinic debates.

Mizrachi's engagement with Jewish law also breaks with its traditional mode of analogical reasoning. By "analogical," scholars mean, for example, that a prohibition against lighting a fire on the Sabbath extends to other kinds of work, or that the prohibition on conquest extends to other uses of force (Fishbane 1998: introd.). In these rabbinic commentaries, similarities and principles derived from key textual passages are sought out from a variety of different cases and linked together by analogy, admitting historical change in the process (ibid.). In other words, considering what is like and unlike across different time periods and among different commentators means that a contemporary engagement with these commentaries requires considerable reflexivity and historicity. But in Mizrachi, this historical dimension of interpretation gives way to a focus on reiteration and consensus, and, therefore, the medieval commentators all seem to be directly sanctioning contemporary practices of settling as carried out in Hebron and the rest of the West Bank. After Mizrachi concludes that a rabbinic consensus on settling actually exists, he then turns to the issue of non-Jews in the land.

In reading present concerns into the past, Mizrachi begins by quoting a key dispute between Rashi and Nachmanides (Ramban), who both comment on the passage "And you shall take possession of the land and dwell in it" [*ve-horashtem et ha-arets vi-yeshavtem bah*] for I have given it to you to possess" (Numbers 33:53). The debate between the two depends on the meaning of the term "possession." Rashi argues that settling in the land is the critical idea, while Nachmanides argues that conquering it alone and taking it back from idol worshippers who are living in the land is enough. The dispute animates Mizrachi and leads him into a consideration of the tractate *'Avodah Zarah* (Idol Worship) in the Talmud, focusing on the well-known prohibition "*Lo tehanem*," or not allowing idol worshippers to dwell in your midst. Mizrachi then quotes Rambam, who says: "We were warned against allowing idol worshippers to reside in our land so that we would not learn their heresies."

Mizrachi says the prohibitions on showing idol worshippers mercy or favor, giving them gifts, selling them land, leasing houses, and extending agreements (peace) to them are indeed still valid today unless they accept the seven Noahide laws, which are the universal laws that are supposed to govern all of human behavior. If they do so, he tells us, their status changes into that of a *ger toshav*, or resident alien, who is granted a series of rights and protections.[29] Citing Rambam again, Mizrachi notes, "if they [idol worshippers] come to terms with and receive the seven commandments of Noah, you will not kill a soul among them, and they must pay a tax, since it says in the Bible 'and they will pay taxes and be slaves.'" Mizrachi then emphasizes that "Arab" residents of the West Bank are *not* idol worshippers and cannot be expelled. Yet he concludes that there is no obligation in Jewish law to allow them to remain. This is because, he alleges, the category of *ger toshav* is only valid once every fifty years, and this Jubilee year is part of messianic law, which doesn't apply to the human era. In sum, he uses the divide between human law and hypothetical messianic law to argue that the rights and protections for "resident aliens," or Palestinians, are only to be applied in the messianic era, legitimating their removal in the present. Mirroring the legal gray zone within the occupation, Mizrachi selects among the religious laws that can be implemented (e.g., taking possession of land) and those that are to be implemented in the future (granting Palestinian rights) to justify settling in the occupation. This interpretive exercise is but one example of the way settler practices on the ground are not only motivated by a particular reading of Jewish tradition but are actually read back into it, remaking core values and classic interpretations that compose its ethical domain.

Conclusion

In this chapter, I have examined three aspects of settler practice—movement, inscription, and boundary formation—critical to creating ethnically distinct settled spaces. Settler versions of Jewishness come to be read in relation to routine trajectories, strongholds, and limits, inscribing potentialities and parameters in the character of Hebron as a sacred place of origin, particularly in the absence of civil and secular law. An ethnic inscription in space, emerging from the abiding logic of these ongoing practices, also becomes key to assuming permanent religious control over Hebron with the overlay of halakhic (Jewish legal) authority. Dissenting views among key rabbinic commentators that challenge existing settler views are either ignored or retroactively amended to conform to an ideological agenda.

Otherworldly messianism, then, may not be the main driving force of ideological settlement as other scholars suggest.[30] According to Gideon Aran (1990, 2013), for example, the secular political realm gets encompassed and reinterpreted as part of a religious worldview, and settlement becomes an "appropriation of Zionism by religion [that] has been made possible by the 'mystification' and 'messianization' of the normative Jewish conception" (1990:162). While appropriations of Zionism are indeed part of the issue, so too are the many changes to Jewish tradition and religious legal argumentation that routinely take place through settler life in militarized and occupied contexts. Focusing on messianic belief alone to the exclusion of a religious settler's preoccupations with treatment of the other as settlers move through and inhabit Palestinian space misses an important way in which settler Judaism is being shaped by its colonial preoccupations. These values have been retroactively read into the tradition, sanctioning force to realize a messianic ideal. Chapter 5 further explores the direction of this religious change, moving beyond what is presumed to be an excess of mystical belief to the core contradictions and perceptions of difference that frequently give rise to settler violence.

Chapter 5

Religious Violence

> We do not believe that anyone can be blamed for not having foreseen the fact that a Jew would plan and carry out a massacre of Moslems in the Tomb of the Patriarchs. Those in charge of security at the Tomb were given no intelligence reports that an attack by a Jew against Moslem worshippers could be expected, particularly since intelligence reports warned of the opposite: an attack by Hamas.
> —Shamgar Commission report, June 26, 1994

Sometime between 4:45 a.m. and 5:20 a.m. on February 24, 1994, Baruch Goldstein, one of Kiryat Arba's three doctors, radioed the settlement's security coordinator and asked to be driven to Meʿarat ha-Makhpelah.[1] This was interpreted by the driver of the jeep, contacted while circling the settlement on guard duty, as a routine request by the doctor on call to be dropped off at the Tomb in order to attend morning prayers. It was the Jewish holiday of Purim, and Goldstein was known to be deeply observant. Yet when the driver picked him up, he was dressed in military uniform with no prayer shawl, as if on reserve duty. At the time, his uniform did not seem unusual given the fact that many religious settlers serve in the military. The jeep was part of a security patrol that operates in Kiryat Arba twenty-four hours a day. This vehicle was linked both to a security center inside Kiryat Arba and to a military command center that coordinated operations in Hebron. The military-like jeep that patrolled Kiryat Arba belonged to the settlement, while the driver, though technically a civilian, was armed with an automatic weapon. The networks that Goldstein made use of to reach the Tomb amounted to more than a simple alliance between settlers and the military. Because the soldiers deployed in the area had long-standing links to settlers living in Kiryat Arba, they supported and participated in aspects of Jewish observance even if not devout themselves. Moreover, soldiers were stationed inside and around the partitioned mosque, ostensibly to oversee the security of both Jewish and Muslim worshippers at the site. The ambiguous state of affairs with respect to the way the civil-military divide operated were key conditions that shaped Goldstein's paradigmatic act of violence.

In this particular year, Purim and the beginning of Ramadan fell on the same day. Under the military's division of space and the rotating arrangements for religious holidays it devised, this meant that Jewish settlers who regularly worshipped in the courtyard of the Tomb and Muslims worshipping in the Hall of Isaac had competing claims to use the entire space of the site. Goldstein was dropped off at the entrance; witnesses from the settlement and soldiers stationed at the Tomb disagree over whether Goldstein entered alone or with accomplices and the actual weapon he was seen carrying. But what is not in doubt is that Goldstein carried an automatic weapon with hundreds of rounds of ammunition into the mosque. His access was virtually guaranteed by the ambiguity of his positioning and the respect he commanded from others he encountered. He was known as a physician, a former member of Kiryat Arba's municipal council, and a commander on reserve duty. Yet within less than half an hour after his arrival, he forced his way through a locked door from the Jewish area of the Tomb into the Hall of Isaac, the main area of Muslim worship, and shot approximately 110 rounds of ammunition into a crowd of nearly eight hundred people worshipping during Ramadan.

One of the many ironies of the incident lies in the fact that it was Goldstein that the security center of Kiryat Arba tried to contact upon learning, via its military communications system, that rounds were being fired in Meʿarat ha-Makhpelah/al-Ḥaram al-Ibrahimi. This center was charged with dispatching ambulances and fire engines as a comprehensive response to emergency medical needs, and it immediately sought to coordinate an extensive rescue effort. Connections between security, military, and emergency medical care had been set up in anticipation of a particular kind of violence— that directed exclusively toward Jewish settlers by Palestinians in Hebron. These emergency arrangements, then, reveal an operationalized notion of violence that anticipates Palestinians targeting a settler presence rather than the reverse.

Many of the basic details of the massacre that took place that dawn will never be fully known. In journalistic accounts that span a five-month period, the figures of the dead and injured on the Palestinian side vary widely. They are inconsistently cited as 29, 39, and 48, while the numbers of injured are variously held to be 150, 200, and 400 people.[2] The discrepancies in these numbers are themselves a telling feature of the incident, particularly since casualties and injuries on the Israeli side in the aftermath of terror attacks are documented in a much more comprehensive way. The lack of value placed on Palestinian lives and of infrastructure to sustain them are both reflected here. Moreover, these inexact figures also point to a significant if overlooked feature of the massacre, which is that the parameters of the "event" remain indeterminate. After the massacre took place, many other worshippers in the mosque incurred injuries in the rush to escape. Israeli soldiers stationed at the site fired shots into the air or at the panicked crowd, which began rushing toward them. Soldiers allege their shots were fired to impose public order and that no other Palestinian deaths occurred as a result.[3] Yet Palestinian accounts dispute this version of events, and many other deaths are said to have occurred from soldiers' firing into the crowd. More deaths and injuries later resulted in clashes that took place during Palestinian protests near a number of West Bank hospitals treating the wounded. Moreover, Palestinians in the mosque who were injured did not cooperate with Israeli military rescue efforts because they did not trust the occupying soldiers. The injured instead relied on the help of friends and acquaintances who carried them out on their own and on family members who rushed to the scene with private vehicles.

Religious Violence

Many scholars have argued that religious violence comes out of a moment of crisis and that the signing of the 1993 Oslo agreements led to a period when settlers in Kiryat Arba felt especially betrayed by the idea that neighboring Palestinian residents would be legally armed. Motti Inbari (2012), Michael Feige (2009), and Ian Lustick (1988) each have a version of a theory of "cognitive dissonance" that spurs devout believers to violence, alleging that when they are confronted with irrefutable evidence of a reality that goes against their beliefs, they reinforce rather than change their views.[4] As Inbari (2012:5) notes, believers "cling more strongly, and try to implement them [their views] by any means possible." Or as Feige (2009:167) put it, there was a huge gulf between "metaphorical Hebron" and "actual Hebron," b Hebron as a "symbolic center" and a "poor endangered peripheral de ment town," that then created an "explosive atmosphere" that fueled violence. These authors see violence as a way of trying to "close the gap" between belief and reality and even "expedite salvation," giving weight to the idea that religious beliefs defy reason and create cognitive dilemmas for believers.

Yet as we have seen in other instances, the tension or gap between what is and what settlers believe ought to be is embedded in the everyday practices and material conditions of settling. It is not as if reality and belief suddenly fall out of line. Rather, they are more often than not out of sync to begin with. For this reason, we need a more nuanced explanation of settler violence, beginning with the observation that the gradual process of remaking the sacred within a militarized context has long relied on considerable uses of violence. Historical events such as Oslo may heighten the stakes and serve as catalysts, but as the Goldstein massacre demonstrates, violence is also shaped by a host of local conditions that directly enable the radical convictions of perpetrators to be implemented.[5] Absent in this "crisis of belief" explanation is a sense of the *longue durée*. For this reason, I explore the massacre in relation to everyday settler practices and ideologies that revolve around ethnicizing and reordering space, this time at a sacred site.[6]

If previously I looked at the ways settler interpretations and practices shifted the emphases of biblical narratives and narrowed the parameters of debate within Jewish legal tradition—reading into the medieval rabbinic interpretation concerns with sovereignty and possession of land alone—here I begin by focusing on the reification of sacred origins in the ideological settler imaginary. I consider the kinds of work settlers have done to remake the materiality of Jewish "origins," as well as elevating the status of the matriarchs and patriarchs to saints in their own right. I argue that the Tomb of the Patriarch's theological power lies in the way its sacredness mirrors, incorporates, and confirms key dimensions of an existing social order, specifically aspects of the occupation and the asymmetrical relations on which it depends. The evidence presented here suggests that a settler version of Jewish sacredness has been shaped as much by incursions, cycles of violence, and the partition of the Ibrahimi Mosque as by particular understandings of Jewish textual tradition. For this reason, the Goldstein massacre, as an act of religious violence, came out of a place-based mode of Jewish observance, stoking anxieties around "difference" and the dangers of "mixing" bound up with it.

The role of ideological settlers in perpetrating acts of violence that target

Palestinian civilians raises questions that pertain to all fundamentalisms—namely, the analytical weight to give to the religious sphere in cases of violence. How might ideological certainty or subjective engagement with the materiality of a sacred site precipitate violence? Is violence caused by the stark divide between self and other that results from a settler way of observing Jewish tradition? The Goldstein massacre is seldom seen as having anything to do with a particular version of Judaism in spite of the fact that the perpetrator was devout, the violence occurred inside an embattled sacred site, the victims were Muslim, and Goldstein's supporters conferred on him a saintly status in the aftermath of the incident. As an analyst, then, I aim to take these aspects of the massacre into account, as well as the incremental remaking of sacred space itself. To paraphrase the historian of religion Jonathan Z. Smith (1982:104), it is often the case that religion's darker sides escape scrutiny because of the dilemma it poses for sorting out issues of tolerance and judgment. As he notes, in Jerusalem, it has proven far more difficult to judge those who advocate building the Third Temple than those who actively have taken measures to destroy the Temple Mount (ibid.), when in fact both deserve scrutiny.

Remaking the Sacred

Official assessments of the massacre do not consider this act of violence in relation to settler property takeovers, the reordering of space, and an emerging radical religious sensibility. The Shamgar Commission report (1994b) focused exclusively on what it deemed to be the criminal aspects of the massacre. It looked primarily at questions of culpability, asking: Were others involved? Who had responsibility? What sorts of checks were in place against violence? Ironically, it skirted the role of religion altogether by examining from a security standpoint some of the technical sides of the site's military arrangements: the number of soldiers on duty, the condition of surveillance cameras, metal detectors, locks, the practice of bringing arms into the site, and the casualness with which the civil-military divide operated. The commission's main conclusion was that Goldstein acted alone and took advantage of his rank and position to gain entrance to the site. It also noted his siege mentality, placing the responsibility for the massacre solely on the perpetrator's deranged psychological state. Its recommendations for the prevention of repeat settler attacks include better locks, a more permanent partition, and adequate security personnel, all measures that maintain the status quo of settling and its devotional links to the Tomb as a key site of Jewish origin. The report did not venture into settler practices that condition attitudes toward the "other" or generate the very hatreds that fuel violence. Nor did it consider the militarization of religion or settler Judaism's narrowing focus on securing exclusive control over what it deemed to be inherited biblical or Jewish property. It defended the military's responses, its ethic of noninterference in settler affairs, and the legitimacy of allowing settlers access to pray at the site. Historians such as Idith Zertal and Akiva Eldar (2009:337), however, allege that many of these commissions and the military-oriented jurist Shamgar himself have a long history of exonerating settler illegalities.[7]

It is difficult to separate out the individual characteristics of the perpetrator from the wider social and religious context that shaped and supported his actions. The Goldstein massacre shared, with other acts of encroachment and transgression, a preoccupation with erasing the very limits that marked the viability of the Palestinian presence in Hebron and elsewhere. The materiality of the sacred space itself, and the centrality of Meʿarat ha-Makhpelah in settler Judaism, played a key role in the violence. In terms of the edifice in which the massacre took place, Jewish settler versions of the "sacred" within it had already featured the imposition of Jewish signs over and against a standing Islamic backdrop. This battle of signs—tactile, aural, and visual—was a key feature of sacredness remade. Profane elements introduced into the sacred space were not, as others have argued (Friedland and Hecht 2000), extrinsic matters that got accidently projected onto the religious sphere but became part of its absorptive and malleable character. Built over a long history of conquest and coexistence, the edifice is a composite of structures that reflect distinct periods of rule, as in Jerusalem. Its massive square stone outer walls are Herodian, but it later served as a mosque tended by Umayyad and Fatimid caliphs. In the interim, an internal dome was built when the site became a church during the Byzantine period. Two distinctive square minarets remain that were added by Saladin when he defeated the crusaders who also took control of the site. Moreover, the structure contains inscriptions and architectural features from the Ayyubid, Mamluk, and Ottoman periods including one of the oldest carved wooden Ayyubid-period minbars (pulpits) still in use. Therefore, the Tomb of the Patriarchs operated as a mosque from the seventh century (apart from its crusader period) until the 1967 Israeli occupation of the area, when it was incrementally divided between incoming radical Jewish settlers and Palestinian Muslim worshippers.[8] Once partitioned, the Tomb became not merely a site of violence but a zone that actively underscored and generated ongoing battles and acts of destruction. Boundaries delimiting areas of control, the unseen other, and the tombs of biblical figures were some of the elements that settlers turned into a grammar of the sacred. After these reinscribed an existing sacred space, ideological settlers further anchored their presence by presenting it as inherited from antiquity rather than fashioned by a contemporary social process.

In 1967, during the initial period of the Israeli occupation, the government opened the Tomb of the Patriarchs for Jewish visitation and worship but tried to set limits on changes that could be made to the status quo of the site. This seeming contradiction of allowing direct Jewish access without changing the identity of the built structure was challenged and transformed by settler actions on the ground. Just as Palestinian vernacular architecture had served as a backdrop for a Jewish settler sense of historicity in other areas, the Islamic character of the site became an important element of its sacred Jewish aura. The site's Islamic features—basins for partial ablutions, or the washing of hands, mouth, nostrils, arms, head, and feet before prayer—then, were resignified and turned into a setting for hand washing and purification before Jewish prayer. This was achieved by placing several double-handled pitchers on top of the basins that had previously been used to purify Muslim worshippers. Similarly, the raised cenotaphs of the patriarchs within the mosque, covered with Ottoman-era green silks embroidered with gold-threaded Koranic suras and located within separate rooms, were designated as part of the new synagogue by affixing Hebrew plaques to their protective grates. *Mezuzot* (biblical inscriptions on doorposts) were similarly placed on all the stone entrances, competing with the Koranic inscriptions above. At the entrance to a smaller chamber leading to the Hall of Isaac on the Muslim side, settlers hung a golden cloth embroidered in Hebrew stating, "To the Hall of the Tombs of Isaac and Rebecca," an area Jewish worshippers do not regularly access. The inscription that follows it comes from Genesis and reads, "Isaac brought Rebecca to the tent of his mother Sarah, and took Rebecca as his wife," invoking the marriage between Isaac and Rebecca. In contrast, ayahs from Surah Saad in the Koran, originally carved into the stone lintels above, read "And remember Isma'il [Ishmael] and al-Yasa'a [Elisha] and Dhul-Kifl [Ezekiel], all among the best. The gates to the Garden of Eden are open to them." The interior courtyard of the mosque has been remade with a canopy and ark holding Torah scrolls. Elements of potential convergence in Judaism and Islam are disassociated and made into competing either/or propositions. This competition is created in part through the material remaking of the site rather than through inherent features of the traditions themselves, since Islam recognizes Judaism's matriarchs and patriarchs as prophets, and Judaism recognizes Ishmael as the brother of Isaac. The rivalries are therefore not preordained in spite of real differences between Jewish figures in the Bible and their rendition in the Koran.

Figure 12. Golden cloth covering the door to the Muslim side of the mosque, 2007. The Hebrew wording on the cloth states: "To the Hall of the Tombs of Isaac and Rebecca"; the hall is an area Jewish worshippers do not regularly access. Biblical markers and Koranic inscriptions in the stone lintels above compete with one another. Photo by author.

Violence at the site emerges from a particular set of social conditions, rather than from the inherent character of these religious traditions or even the space of the sacred itself. Against constructivist conclusions such as these, however, Ron Hassner posits a different explanation. He asserts that sharing the sacred always elicits violence because the sacred is by definition "indivisible." As a "conduit" to the divine, he argues, the sacred is exclusive and cannot be split among different groups of believers (2013:3).[9] Moreover, he does not put much emphasis, as I do here, on remaking sacred spaces, because he sees existing "social facts" as a constraint on making changes to them; "the historical and social weight" of sacred sites grants these locations a sense of permanence that can be "difficult and costly" to transform (ibid., 10). Moreover, Hassner maintains that if sacred sites are those central to a tradition, the chances that dividing them will elicit violence increase exponentially. While these key principles are intended to outline the universal characteristics of religious conflict, they minimize the agency of believers in creating the conditions for violence in particular contexts, and the way ideological groups may actually reconfigure sacred sites as anchors for other social spheres.[10] In Hebron, for instance, it is not sacred space alone that is at issue but armed Jewish settler forays into Palestinian space that leave behind the signs and effects of erasure and destruction.

Taking account of settler agency as it operates within the conditions of an occupation in contrast to Hassner, then, I emphasize the ways settlers have incrementally remade the sacred and the violence this has entailed. That the sacred is mediated through a set of contemporary conditions may depart from the theological views of Jewish settler worshippers themselves. Yet devout settlers ironically acknowledge the Tomb's contradictions and anomalous features as outlined below. Unlike in Christianity or Islam, any synagogue that has been built over tombs presents a host of purification issues for Jewish observance. There are extensive injunctions in Jewish law against worshipping near corpses because they are considered impure, and therefore most concede that Torah scrolls should not be placed near graves. When I inquired about this, the purity dilemma wasn't dismissed out of hand, but it was generally circumvented. Rabbis in the settlement and ideological settlers asserted that the graves of Jewish ancestors were actually located far beneath the ground. Yet, this did not address the issue of the raised tombs directly within the space settlers claimed as a synagogue. Jewish settlers did not maintain that these tombs were empty, in other words, even if they believed they did not contain the actual remains of the patriarchs themselves, and so the area operating as a synagogue was believed to contain graves within it. Some settlers acknowledged that for this very reason, those descended from the Jewish priestly class (e.g., Kohanim and Leviim) did not pray at this synagogue.

Another contradiction then followed: if those praying at the site believed that the tombs above ground were not the true graves of Jewish patriarchs and matriarchs, then why did they insist on direct access to them? A syncretic form of Jewish tradition shaped in relation to Christianity and Islam in the Middle East has led to saint worship, but this is not a formally accepted practice in the European Ashkenazi version of observance practiced by most Jewish settlers. Within the Tomb itself, devout settlers of all types treat Judaism's patriarchs and matriarchs not only as historical figures but as saints. Their sacredness is both historicized and materialized in ways that are not characteristic of other possible modes of Jewish worship. Several settlers told me, for instance, that they asked for the intercession of Abraham, Sarah, Isaac, Rebecca, and Leah in their daily lives, treating these figures as intermediaries or gods in their own right. This appears to be a departure from monotheism and a curious remaking of Judaism's foundational injunction against idolatry. The settler insistence on praying directly inside the Tomb and at the side of individual tombs, then, created inconsistencies from within the terms of existing Jewish belief and practice, as did the push to access the Muslim side of the mosque to pray directly beside the Tomb of Isaac. What it reveals, however, apart from changing forms of religious practice, is a contemporary preoccupation with securing a greater place for Judaism over and against a competing Islamic presence. It is also not verifiable that the graves in Hebron were regular sites of Jewish pilgrimage as was the Jerusalem Temple. The scholar Jacob Neusner (personal communication) noted: "When Jews went to graves, it was not for pilgrimage but to bury the bones [of the deceased]."[11] The evidence favors the idea that the specific ways Hebron's Jewish sacredness is being observed today is as much a contemporary engagement with the past as it is about simply carrying out a long-standing tradition.[12] The graves of the patriarchs and matriarchs, in other words, as well as Jewish worship directly within the space of the mosque itself, have been given a new centrality that competes with traditional Jewish concerns for the destroyed Temple in Jerusalem. It is this contemporary set of choices over marking Hebron's Jewish sacredness that reveals a devotional practice shaped as much in relation to the logic of settlement as to received religious tradition. This is not to charge contemporary Jewish observance with a lack of authenticity but rather to suggest that the choices that have been made with respect to the Tomb are those that have foreclosed other opportunities for peaceful coexistence between Jews and Muslims.

Partitioning the Sacred Site

The role of Jewish settlers in the gradual partitioning of the mosque is clearly evident in its microhistory as discussed in the Shamgar report. The report reveals how the extension of a Jewish settler presence in the Tomb, a monumental structure that, as previously noted, had served as a mosque from the seventh century until the Israeli occupation,[13] transformed it into an arena of confrontation long before the Goldstein massacre ever took place. For Hebron's Jewish settlers, the government's regulations around the sharing of the space became artificially imposed limits that needed to be modified according to their understanding of Jewish tradition and authenticity. Muslim worshippers, on the other hand, considered the Jewish settler presence inside the mosque to be part of a larger process of colonization. A triangulating struggle to control the site between Jewish settlers, Muslim worshippers, and the Israeli state led to a series of confrontations that prefigured the Goldstein massacre. Israeli soldiers and the military police stationed in the mosque to maintain order were structurally allied with religious settlers—which meant that their concern with allowing equal access to Jews and Muslims at the site emboldened Jewish settlers. One of the consequences of this asymmetrical order was that the military was easily drawn into Goldstein's paramilitary activity.[14] The Israeli government's contention that the soldiers stationed at the Tomb were mainly there to ensure "religious pluralism" and the security of "both communities" instantly fell apart.[15]

The chronology of government resolutions documented in the Shamgar report reveals a series of accommodations to settlers that profoundly changed the character of the mosque in a matter of four years. While initially Jewish settlers were only allowed to visit and pray at its entrance, the government reversed its policy, and the site was divided using a low movable barrier that let both religious groups pray inside it, while seeing and hearing one another. Over time the barrier was replaced with a more permanent wall, but the aural competition around Jewish versus Muslim prayers continued and still exists to this day.[16] From the time the government first allowed Jewish visitation and prayer at the Tomb, the original settler nucleus of Hebron, which established Kiryat Arba, agitated for greater concessions. They began a campaign of letter writing to keep the issue of Jewish access prominently in the public

Figure 13. The security entrance to the Jewish side of the Tomb of the Patriarchs, 1996. There are separate metal detectors for observant men and women. The sign on the left, with an arrow, reads, "Deposit for Weapons." Photo by author.

domain, while at the same time generating a steady stream of official correspondence directly addressed to Moshe Dayan, the defense minister and the military governor of the West Bank. They presented themselves as devout Jews rather than religious right settlers encumbered by heavily restricted access to a site sacred to all Jews. Moreover, they challenged the site's identity as a mosque by invoking biblical passages referring to Abraham's purchase and his full ownership of the land beneath it as the evidentiary basis for their claims. A portion of the Shamgar Commission report (1994b:99) presents the government's responses to these appeals emphasizing both permissions and several limits (italics added):

> November 15, 1967—In the framework of a ministerial committee, an appeal was discussed regarding the permit Defense Minister Moshe Dayan gave to open the Meʿarat ha-Makhpelah to Jews on Saturdays.

> July 24, 1968—Defense Minister Moshe Dayan mentioned that the entrance to the Meʿarat ha-Makhpelah is only for prayer and visiting, *and that weddings may not be held either in Meʿarat ha-Makhpelah or next to it.*
>
> October 6, 1968—It was determined that a unit of the Military Rabbinate will continue to supervise the visiting arrangements and public order at the Meʿarat ha-Makhpelah. On days when tension or large crowds are expected, a unit of the Military Police will be stationed at the Meʿarat ha-Makhpelah to secure order there. Toward Yom Kippur prayers the Defense Minister will hold a meeting with Hebron dignitaries in order to inform them about the prayer arrangements on this holiday at the Meʿarat ha-Makhpelah. *It was resolved not to view the Meʿarat ha-Makhpelah as a synagogue.*

This series of resolutions in response to settler appeals for greater access highlights the contradiction of the government's visitation policy: generally limiting Jewish access to visiting the site while nevertheless facilitating large Jewish gatherings there on the high holidays and maintaining that the site was a mosque. As is clear from these resolutions, the government initially prohibited Jewish use of the mosque for anything other than prayer one day a week—the very limits that settlers militated against. Disputes between the Hebron settlers and the police erupted over the practice of holding secret weddings and illegal circumcisions within the mosque because they introduced food, beverages and alcohol desecrating the sacred character of Muslim space. This process of encroachment in the context of the occupation sparked violent retaliations:

> October 9, 1968—A grenade was thrown at the outer stairs leading to the Meʿarat ha-Makhpelah and 43 Jews were wounded, most of them women and children, among them five severely. Following this, it was decided that visits to the Cave, although they should continue in their proper order, should be conducted with increased protection for the visitors. It was also decided to place soldiers in the houses in the vicinity of the Meʿarat ha-Makhpelah. (ibid.)

The injury of devout Jewish women and children played into the hands of ideological settlers by permanently increasing the number of soldiers stationed in the Palestinian residential areas adjacent to the mosque in order to ensure their protection. Celebrations on the Jewish settler side were subsequently brought within the mosque, and settlers made full use of the many immobile elements of prayer and ritual to establish their permanent presence. Therefore, not only did settlers pray at the mosque more frequently, but they began introducing accoutrements of prayer (arks, Torah scrolls, prayer books) and furniture (chairs, bookcases, and dividing walls) to permanently claim the space.[17] Muslim worshippers countered by holding funerals within the entire area to desecrate Jewish notions of purity.

By 1975 in light of growing confrontations, the government decided to remake the spatial and temporal boundaries to permanently accommodate a Jewish settler presence inside the mosque. The harassment and unrest sparked by this expanding settler presence ultimately led to the mosque's initial partition:

> August 4, 1975—Following damage to Jewish religious paraphernalia (which was reported on 27 July 1975) as well as irregularities of Jewish worshippers who brought beverages into the Cave, held weddings and circumcisions in it, ate and smoked, Defense Minister Shimon Peres, and Major General Raphael Vardi gave a review of the status quo as regards the prayer regulations as they evolved after 1967. (ibid., 101)

The review of the status quo resulted in reapportioning the space as follows: The Hall of Abraham and Sarah, the Hall of Jacob and Leah, and the balcony and courtyard connecting these two halls were allotted to the Jewish side, while the Hall of Isaac and Rebecca, the Djaouliyeh Mosque (Hall), and the Yusufiyya Hall continued to remain Muslim. Yet even though settlers had been given more space, it did not put an end to their quest to expand. On the contrary, it led them to try to gain direct access to all of the tombs and rooms containing biblical figures deemed sacred to Jews. Yusufiyya Hall and the Hall of Isaac on the Muslim side therefore became a touchstone for violence. For its part, the government justified the need for expansion not on sacred grounds as did settlers but on the practical need to accommodate the internal diversity of the newly established settlement of Kiryat Arba, and different styles of prayer—Ashkenazi, Sephardic, and Yemenite—of its increasingly diverse Jewish population (see Figure 14 below for a floor plan of the Ibrahimi Mosque).

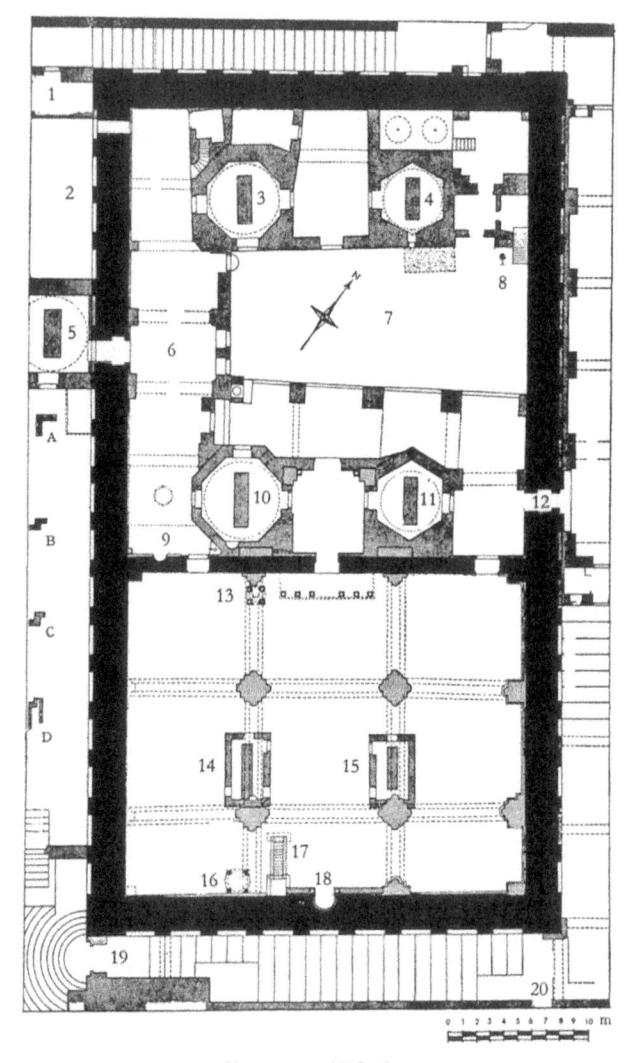

Plan of the Masdjid Ibrāhīm

A. Herodian structure. — B. Byzantine structure. — C. Ancient structure. — D. Arab structure.

1. Entrance. — 2. Joseph's Mosque. — 3. Jacob's tomb. — 4. Leah's tomb. — 5. Joseph's tomb. — 6. Women's Mosque. — 7. Inner court yard. — 8. Date palm. — 9. Adam's footprint. — 10. Abraham's tomb. — 11. Sarah's tomb. — 12. Entrance. — 13. Canopy (opening in floor). — 14. Isaac's tomb. — 15. Rebecca's tomb. — 16. Canopy (sealed opening). — 17. Minbar. — 18. Miḥrāb. — 19. Main entrance. — 20. Entrance.

[After Vincent, *Hébron*].

Figure 14. The floor plan of the Ibrahimi Mosque. See *Encyclopedia of Islam*, 2nd ed., p. 959, s.v. "al-Khalīl." Reproduced by permission of Brill Academic Publishers from *The Encyclopaedia of Islam*, 2nd ed., vol. 4, edited by E. Van Donzel, B. Lewis, and Ch. Pellat (1997).

Layered over this division of space made in the mid-1970s was a temporal rotation meant to accommodate the cycle of key religious holidays for each group. On Yom Kippur or Ramadan, either Jews or Muslims were awarded exclusive access to the entire edifice by the government, again framed as a matter of ensuring the space's "pluralistic" character. In language and tone that resembles the directives of the British Mandate, the Israeli government decision reads: "This announcement [of the domains and times of worship] should be conveyed to the qadis in Hebron, those in charge of the Me'arat ha-Makhpelah, and afterward to the mayor of Hebron and the municipal council of Kiryat Arba. *It should be emphasized in their presence that there is no intention to divide Me'arat ha-Makhpelah, but rather to divide the times of worship and access paths so that everyone can pray to his god and visit the tombs of his saints with mutual respect and without any disturbances*" (Shamgar 1994b:103). Both the qadis (religious judges) and Palestinian mayor of Hebron, however, were subject to the military's authority in ways that the elected council of Kiryat Arba and the municipality were not. Moreover, at the moment of regulating worship, the military ensured a form of Jewish access that masked the settler quest for increasing control and dominion of the site. In 1979, long after the mosque had been used in this rotating manner, another government concession to settlers allowed Jewish worshippers to regularly enter the Muslim side in order to access the tombs of Isaac and Rebecca, with the provision that Jewish worshippers be respectful and "not step on the carpets covering the area that was sacred to the Muslims" (ibid.,104). They were then allowed to pray directly in the Hall of Isaac, but only until 10:30 a.m. because Muslim worship would resume at 11:00 a.m.[18] While these new regulations around sharing were elaborately constructed, they left matters open in the case of overlapping Jewish and Muslim holidays, when "special arrangements would be determined." The indeterminacy around overlapping holidays meant that Jewish settlers in general and Goldstein in particular could take this occasion as an opportunity to further establish precedents on the ground through violence.[19]

On Jewish Versus Secular Law

Even though the military rules and regulations governing the mosque favored ideological settlers, they did not see themselves as bound by the many regulations governing the Tomb, particularly when it meant concessions on matters of Jewish observance. This overall disregard for the mosque's sharing regulations by radical right settlers like Goldstein was part of a larger trend toward lawlessness and the ideological view that they were only truly bound by religious imperatives. If law-abiding political participation and violence stand at opposite ends of the same spectrum (cf. Arendt 1970), the settler sense of not being bound by secular law then opened the door to using violence in the service of realizing higher religious principles and aims. Not only did settlers seek to take over more of the mosque than they were allotted, they also sought to usurp the military's authority to establish rules and regulations governing Jewish access to a sacred space. Here we revisit the issue of legality from another vantage, not just that of the legal gray zone of the occupation that exists but also using a presumably divine mandate to undermine the comparatively secular authority of the state. Goldstein used violence to overturn existing regulations and to attempt to secure full control over the space of the Tomb. In the process, he also challenged the legitimacy of the state and the military, attempting to replace official divisions and rotations over sacred space with greater Jewish (settler) control based on what he took to be a religious requisite to surpass these rules.

Jewish witnesses in the Tomb on the day of the massacre allege that Goldstein became angered by the locked doors leading to the Hall of Isaac preventing his access to the whole mosque on Purim. His earlier infractions of the law, for which three known police files in the Hebron precinct were opened but never pursued due to a "lack of public interest" (ḥoser 'inyan la-tsibur),[20] illuminate the lapses in law enforcement that led to the massacre: On February 7, 1990, Goldstein was arrested because he ripped up a general order prohibiting the entrance of non-Muslims to the Tomb on a Muslim holiday. On December 4, 1991, he was also arrested on suspicion of overturning a bookcase containing copies of the Koran in the Hall of Isaac, allegedly in response to damage to sacred Jewish objects that occurred the day before (Shamgar Commission 1994b:77). These prior incidents of lawlessness demonstrate his disdain for the rules and regulations governing the site, as well as the perils of not prosecuting settlers who repeatedly violate the law. What is less understood, however, is the way his transgressions were viewed in wider settler circles—not as violence directly mandated from above but rather as a legitimate way of bypassing secular limits when laws became major obstacles to realizing religious aims.

As an example of this ambivalent attitude toward the law, I turn to statements Goldstein made in the course of his political career, particularly when running for office as a member of the militant religious Kach Party[21] in both national and municipal elections. In 1984, Goldstein was the third candidate on the Kach list, falling just short of a seat when Rabbi Meir Kahane, the party's leader, was first elected to the Knesset. By 1988, however, the Kach Party was entirely banned from national elections because of its antidemocratic and overtly racist platform. It nevertheless continued to have a local following in economically struggling municipalities like Kiryat Arba where Kahane himself lived for a time. Goldstein, Kahane's acolyte, subsequently ran for office in Kiryat Arba's municipal elections and was elected to office for several terms until he stepped down in 1993, only a year before he carried out the massacre.

In an excerpt from a longer interview distributed as part of his Kach campaign literature prior to these elections in Kiryat Arba, Goldstein emphasized the "relative" rather than absolute obligation to abide by the law as a step toward making the general case that secular law (and the formal authority of the state) needed to be subordinated to religious law. Consider, for example, the assertions he made on preventing "Arab" vehicles from entering the grounds of Kiryat Arba:

> A few days ago, a boy that crossed the road on Shabbat was run over by an Arab driver. His face was bruised and his leg was broken. We have to find the legal means to limit the entrance of Arab vehicles to Kiryat Arba. The law is only a tool for the benefit of citizens, and when the situation on the ground changes, as when we are in an actual war, the law must be adapted to the good of the residents. I am not prepared, as my predecessor was, to live with the assertion that limiting the entrance of foreign vehicles is against the law; if the law does not suit the best interests of the public, it must be changed. (Ben-Horin 1995:347)

Some elements of a standard Kach Party position can be seen in this political statement. Coming out of the streets of Brooklyn initially, Kach saw itself as standing for the defense of vulnerable Jews, as well as enhancing Jewish pride and religious identity. It also claimed to be protecting Jews from "attacks and defilement by non-Jews" and emphasized the dangers of mixing with foreigners, specifically in this case "Arabs." Kach therefore advocated the complete separation between Jews and non-Jews and, more forcefully, the "expulsion of Arabs" from all of Israel and the Occupied Territories. Kahane also became obsessed with sexual relations between Jewish women and Arab men and proposed criminalizing them when he served in the Knesset.

Goldstein carried on the Kach tradition of not condemning violence outright, even in the case of the Jewish Underground, the paramilitary group whose members were sentenced to prison for plotting to destroy the Dome of the Rock and stockpiling weapons for this purpose. In elections, Goldstein advocated the need to change the law, using the language of liberal democracy and the right of the people to determine the best laws for self-rule while channeling these democratic ideals toward purely exclusionary aims and practices. For Kahane as well as Goldstein, these antidemocratic convictions were elevated to religious principles; both figures were convinced that the ultimate goal of a Jewish state was to be governed by Torah alone, and they hoped to see all legal authority shift from Israel's High Court to a resurrected rabbinic court (Mergui and Simonnot 1987:32). In assessing the "extremism" of Kahane's ideological vision, then, one might distinguish between glorifying violence outright and the lack of any practical or phased means for implementing the religious ends he advocated and envisioned. This together with a stated urgency of being under siege or on the brink of war created convictions among Kahane's many thuggish followers, Goldstein included, that violence was the only way to protect the sanctity of Jewish life.

Immediately after the massacre, Goldstein went from being a highly respected figure within settler circles to a disavowed perpetrator of violence, even in the opinion of many residents in Kiryat Arba. Yet by the following year, influential leaders in the settlement, as well as its foremost rabbi, Rav Dov Lior, tried to rescue Kiryat Arba's fallen reputation and salvage Kach's ideas, calling Goldstein's notoriety into question. Political figures like Elyakim Haetzni, a former Knesset member in the defunct far-right religious Teḥiyah Party and founder of Kiryat Arba followed suit. He did not repudiate the ideas of Meir Kahane or Goldstein outright. Rather, he chose to couch his own far-right views as a disagreement over means rather than aims. During a discussion with me about Teḥiyah politics, a party that melded secular and religious rationales for its maximalist position around territory and its refusal to consider any exchanges of it in peace negotiations, Haetzni emphatically said, "Kahane was a brilliant man, a brilliant man—it is a *pity* that we lost such a brilliant leader; we lost him twice, first because of the methods he used and secondly when he was murdered. He marketed his ideas by such brutal means, means that the public could not accept, by such brutal rhetoric, while consulting people who had an image of being rough, and brutal, that he sentenced himself to a tinier slice of the population than he could have achieved if had he chosen other means." His remarks presented Kahane as one who had sound views and (as a victim rather than perpetrator of violence), as one who merely needed to articulate them in a more palatable way. Moreover, Haetzni's own statements reflected the attempt to mainstream such far-right ideas by emphasizing victimization rather than overt uses of violence, as became evident in the rhetoric of his public speeches.

In a May 1996 Jerusalem rally that I recorded, for example, which protested the prospect of an Israeli military withdrawal from Hebron, Haetzni gave expression to the idea that the Jews of Hebron were not bound by secular law or any international agreements, providing the cover of legitimacy for acts of settler violence that would follow. Casting settlers as the true victims of Palestinian terror while sidestepping the lawlessness enabling settler violence, he advocated disobeying the law outright. The irony was that his call to disobey the law in the name of an allegiance to divine authority seemed to his religious settler supporters perfectly compatible with his being a lawyer by profession and his stance as a nonobservant Jew. In a speech before hundreds of ideological settlers protesting concerns about the worsening security situation that Haetzni attributed to the impending military pullout, he remarked:

> They say "what are you complaining about, it's written in the Oslo Accords that Hebron is to be given over," so we want to say in the name of this audience and in the name of tens of thousands, and hundreds of thousands, and millions of Jews in the world, we are loyal to the covenant with God, we are not loyal to Oslo. We don't recognize Oslo. In our eyes Oslo is repealed and canceled and we will rip it to shreds. Oslo is not irreversible, it is now already reversed, and if there are political parties in Israel which need for instrumental reasons and for the sake of elections—and I don't judge them—not to recognize the reality that existed before Olso, let us allow ourselves, and I am not a rabbi but the rabbis will forgive me, to release them from their vow of recognition of Oslo both in my name and in the name of this holy congregation. Any obligation toward Oslo does not apply.

This aspect of Haetzni's speech relies on the performative strategy of bringing a truth into being by declaring it so and by invoking rabbinical authority to presumably cancel the ideological settler community's obligations to recognize Oslo and abide by Israeli and international law. It also expresses the idea that national agreements and laws are not valid in comparison with divine covenants, authorizing settlers to take matters into their own hands. Haetzni clearly states, Oslo is "repealed and canceled" (*batel u-mevutal*) invoking the will of the congregation. Moreover, in his speech, he and the several hundred Jewish settlers speak in the name of "millions" of Jews around the world. The speech ends with a gloomy prophetic moment in which Haetzni asserts that any existing regime contradicting the divine imperative of having Jews remain in Hebron is destined to fall from power: "whichever regime raises its hand against Hebron, that very regime which strikes against Hebron, will fall from power and never again return" (*mi she-merim et yado 'al Ḥevron, oto shilton she-merim et yado 'al Ḥevron, 'al Ḥevron yipol meha-shilton ve-lo yaḥzor le-'olam*). He presents himself as a prophet, and the efficacy of this speech is built on elevating religious authority over secular law. Like those in the Kach Party, Haeztni too emphasizes crisis and a state of abnormal affairs. He also taps into the sense of vulnerability and victimization among settlers. That sense of being directly vulnerable, a sentiment still felt to this day, is paradoxical given that settlers are fully armed and under extensive military protection. Nevertheless, it continues to be an influential rhetoric that effectively mobilizes far-right opposition, and, for this reason, I explore its significance further.

On Threat and Vulnerability

Many settlers and their sympathizers accounted for the Goldstein massacre in terms of "threat" and their own vulnerability. Chaim Simons (1995), a devout Kiryat Arba settler and resident historian, contended the massacre was prompted by reports of an impending attack.[22] The ominous signs of the "threat" he detailed come out of a collective preoccupation with being targeted by Palestinian "enemies." Hamas, Simons maintained, disseminated a pamphlet warning of a planned attack two weeks prior to the Goldstein massacre, and nine officers in the defense establishment, including the mayor of the municipal council of Kiryat Arba, had been alerted to the possibility of this attack. Although his representation of the threat sounds dire, such warnings are apparently not a unique occurrence. Soldiers deployed in the area have mentioned that threat warnings are issued throughout the year and that there is no direct correlation between receiving warnings of this sort and attacks that do occur. The idea that Goldstein was responding to a specific threat makes sense only as a retroactive explanation. This is further indicated by the fact that Goldstein targeted worshippers who were praying at the time; they could not have posed any specific threat to him or others—and, therefore, the logic of this defense breaks down entirely. What is revealing, however, is the framing of Goldstein's actions as defensive and the alleged signs and symptoms of threat that settlers point to, which are so vague that they can only be understood as expressing their fears around perceptions of "difference" itself.

First, Simons (1995) noted that there were "gatherings" in Palestinian areas of Hebron, seeing those purchasing food in the market and coming to pray in the mosque as threatening in spite of the fact this occurred just prior to Ramadan. He also noted that other settlers claimed weapons were being stockpiled in one of the mosque's storerooms, even though that stockpile never materialized.[23] Still others, he asserted, alleged that Arabs were shouting threats like "slaughter the Jews" over the loud speakers, and that this was the same cry heard just prior to the 1929 riots in which sixty-seven members of the Old Hebron Jewish community were killed.[24] While these explanations reveal a good deal about settler anxieties, they tell us little about what shaped their experience of being under siege.

I begin to explore this prevailing sense of vulnerability and victimization here by noting that a number of injuries did occur during part of my fieldwork. Settlers were targeted, and often it was not the most militant settlers who paid a heavy price for the asymmetries of a colonial framework. In the span of the first year of my fieldwork, for instance, the following incidents took place: two nonfatal stabbings of older residents; the death of a young couple in a drive-by shooting (leaving behind their two young children); and two incidents of shootings at the public bus, resulting in one fatality and several injuries. In addition, several windshields were smashed by stone throwers. Compared to the violence directed at Palestinians by armed settlers and soldiers in Hebron, these injuries to the quasi-civilian settler presence were relatively limited, though genuinely intimidating. This was enough to give form to a full-blown discourse around threat, which, according to settlers, posed a grave danger to all of their lives. This sense of threat also served to justify aggressive behavior, creating a stark divide between "us" and "them," while attributing to all Palestinians the potential for violence.[25]

In the wake of the Olso Accords, however, mainstream Israeli politicians were also seen as equally threatening. The Goldstein massacre preceded the assassination of Israel's prime minister Yitzhak Rabin by a year, and within

Kiryat Arba these two incidents were often considered to be parts of a whole (cf. Feige 2016). After the Rabin assassination by Yigal Amir, a religious law student, these intimations of threat among settlers mushroomed into delusions of an outright government conspiracy. Kiryat Arba settlers alleged, for instance, that Avishai Raviv, the leader of Eyal, a Kahanist group operating in Kiryat Arba, had been set up as an informant by the Israeli General Security Service (GSS), or Shabak, the intelligence-gathering apparatus of the state that was working against them.[26] Raviv had apparently lived in Kiryat Arba for two years in a rented apartment, directly above Baruch Goldstein, while presumably on the GSS payroll (see "Ex-Undercover Agent" 1999). If it was true that Raviv had befriended Yigal during this time, settlers alleged, this was only to instigate the Rabin assassination while placing the blame on Kiryat Arba. In other words, many in Kiryat Arba contended that the assassination was a plot devised by the government to further discredit the integrity of those living in the settlement. According to this paranoid view, settlers were being scapegoated for a plot devised by Rabin's own security apparatus. Bits and pieces of evidence for this position were brought together in the most fantastic ways. Some found confirmation of a conspiracy in a video of the Rabin assassination shown on television. The video putatively focused on the assassin, Amir, for several seconds before panning toward the crowd. When the cameraman of the state-owned television channel was interviewed and asked why he filmed what he did, he remarked that he had a strange feeling something was about to happen. For some skeptics like Miriam, a Kiryat Arba settler who was glued to the television as were others, this statement alone was proof that the cameraman had been tipped off about the whole affair.[27] She doubted that the religious radicalism attributed to Kiryat Arba existed at all, and commented that when the media secured a video of an Eyal children's camp showing masked youths in combat training in a hidden location in Kiryat Arba, this was a fake portrayal staged by GSS (Shabak). In her mind, the image didn't fairly represent the mainstream views of Kiryat Arba settlers: it just isn't "representative of our public" (*lo mat'im la-tsibur shelanu*), she asserted.[28] Like others, she minimized the Kahanist influence in Kiryat Arba and mentioned that the very small numbers of Kahanists who actually did live in the settlement were just talk and no action.

Miriam's allegations point to a dismissal of religious radical views among ideological settlers and a willingness to downplay violence, coupled with a deeply felt distrust of the media and the state. Yet, in terms of "threat," they also raise a number of issues bearing on government surveillance. Just after the 1984 discovery of the radical Jewish Underground's plot to bomb five Palestinian buses with many passengers, the Israeli government responded to what they considered to be a growing settler subculture of violent extremism by extensively monitoring those living in Kiryat Arba. Settlers claimed the government had attempted to infiltrate their communities from within and to keep tabs on any potential criminal activities by putting resident informers on its payroll. Allegedly these surveillance activities were heightened after the Goldstein massacre, and from the vantage of settler opinion they were even further expanded after the Rabin assassination. During that period, I was told that people only really trusted their longtime friends, and that the original "nucleus" (gar'in) of settler families trusted themselves alone and no one else. Residents contended that their phones had been tapped, rooms bugged, and that informers had even slipped into religious services only to report back on different factions within the community. Some of these allegations may have been true, but they also point to the denial of any radical presence bound up with Kiryat Arba's growing propensity for violence.

Memorializing as Violence

The defensiveness around Goldstein's use of violence among settlers in Kiryat Arba gave rise to his remaking as a hero and saint after his death. The attitudes of many settlers were characterized by denial, racism, and an unwillingness to reflect on the contradictions inherent in the sorts of ideological commitments that had led to the massacre in the first place. Goldstein had been a respected resident of Kiryat Arba, and both neighbors and acquaintances were often unwilling to express any criticism of him. Their overall identification with Goldstein may have been more ambivalent than that reported in the media, but many positive attitudes were expressed to me. For instance, Eddie, a U.S. immigrant and veteran of the Korean War living in Kiryat Arba, who had hung a portrait of Goldstein on his wall after the massacre, said that Goldstein was "an aristocrat, someone way above himself" and that he missed him. He bragged that, as a "munitions expert," he had taught Goldstein to shoot, and then added that Goldstein "shouldn't have done it" because his life was "worth way too much." His statement of reverence for Goldstein was indicative of the way Kiryat Arba settlers valued the life of a single Jewish perpetrator over the lives of the many Palestinians he had killed and injured.

Another instance of denial involved Goldstein's wife, who according to a Palestinian activist, had the gall to continue pushing for a further investigation into those who had killed her husband, while demanding both compensation for his death and that those responsible be brought to justice. Goldstein died at the hands of worshippers in the Hall of Isaac who disarmed and beat him. Yet in the eyes of his immediate family, he was murdered. Their claim is one of many examples of inverting power relations and reframing perpetrators as victims. While some settlers claimed that Goldstein was their friend and neighbor, many others downplayed his violence as a way of expressing their loyalty to Kiryat Arba in the face of Israeli and international condemnation. Shulamit, a Jewish convert with misgivings about the massacre, for example, related the following: "On hearing that my daughter [living in the Jewish enclave in Hebron] defended Goldstein, I went to visit her and say, 'Murder is murder—what Baruch Goldstein did was none other than murder— killing people at prayer. These are not the values I raised you with.' But when the entire world turned against us in the days after the massacre, I decided to attend his funeral to show solidarity."

Others expressed a similar ambivalence. Their complicity with violence involved downplaying the massacre in comparison with Goldstein's favorable deeds as a doctor. Menachem, another informant, mentioned that he remembered Goldstein and liked him well and that his main mistake was that he "hated his enemies too much" and took "God's decisions" into his own hands. Though noting his disagreement with Goldstein's use of violence, he nevertheless asserted that he could not judge him and then told me this anecdote: "Arriving for reserve duty, I was grabbed by some of my fellow reservists and asked to tell them whether I thought Baruch Goldstein was in heaven or hell. I told them 'I can't judge a person in their absence, much less judge the dead, because that would be a stain on my own character.'" This assertion says a good deal about the way "not judging" takes precedence over legal and religious injunctions against murder. While attitudes such as this reveal an abiding complicity with violence, they do not idealize Goldstein in the way many eulogies of his life did in the aftermath of the massacre.[29] Zehava Nativ (Ben-Horin 1995:398), for instance, glorified Goldstein in her recollection of the way he presumably saved the life of a premature infant born in an ambulance. When an ambulance was about to crash into a chain, she wrote, Goldstein "turned around, faced the rear, and protected the infant with all of his body, saving the baby's life." The maternalist theme reemerges here, and this time it recuperates Goldstein as a hero: "The great physician knew that if the tiny baby (who was breathing with difficulty) would have absorbed the blow, he would have been crushed" (ibid.). Trying to shape this incident into a moral parable, Nativ then concluded by asking: "Is Goldstein not a shock absorber for the nation?"

Shrines and Saint Worship

These attempts at memorialization are more than an individual's idiosyncratic views—they express a collective ethos that supports settler violence. Goldstein's memorialization, more than anything else, reveals the degree to which violence is deeply embedded in the ideological settler project. As further evidence for this, I turn to the conflict around Goldstein's grave, which became a shrine to his actions and led others to venerate him as a saint. While residents of Kiryat Arba insisted that Goldstein's burial site within the settlement was chosen by default, it became an integral part of a growing network of saints and martyrs in Jewish Hebron. Though set apart, it symbolically and spatially was linked to the patriarchs and matriarchs in the Tomb, the graves of those who had died natural or untimely deaths in the settlement, as well as those who had been massacred in the riots of 1929.

The Israeli military did not allow Goldstein's family to bury the body in Jerusalem or in Hebron's cemetery out of concern that protecting the grave would create an additional security burden. Instead, the perpetrator's body was buried within the limits of Kiryat Arba itself. A settler living in Kiryat Arba alleged that her son-in-law drove the body from place to place in search of a suitable burial site and that Goldstein was buried in a common area within the settlement, dedicated to the memory of Meir Kahane, only as a last resort. This gravesite seemed abandoned during the early part of my fieldwork. Once I noticed two teenage girls praying there. There was rumor, however, that groups or individuals did come to visit the area, either in protest or support of Goldstein. A busload of Sephardi women was alleged to have stopped at the grave to pray as though Goldstein were a revered saint. It also was rumored that a protester from Tel Aviv came there the day after the Rabin assassination and splashed the area with black paint and that a soldier serving in Hebron committed suicide there. The park also had a reputation as a place of prostitution.

It was, ironically, the grave's location inside the settlement that helped generate the conditions for Goldstein's resurrection as a saint. His raised grave was placed on paving stones in a plaza, and benches were set up to accommodate visitors. Another stone shelf held prayer books, while gas lamps lit the area at night. The site was contradictory. On the one hand, it seemed as if these were attempts to "normalize" a tomb out of place. Yet the gravesite also had the feel of a religious memorial, and the inscription on the tombstone invited onlookers to participate in Goldstein's glorification. Referring to him as a "sanctified physician" and "rabbi" who "gave his life for the nation of Israel" and as a man "killed in the service of God," it whitewashed the massacre. Like other settler attempts at resignifying the mosque, memorializing the dead, and putting together an array of historical and biblical events to form a continuous whole, the grave too became part of an ongoing process of material inscription that lies at the heart of settlement as an ideological process. These all serve to create permanent links between Jewish settler sites of origin, memory, and, as this case demonstrates, burial and death.[30]

It was this romanticized hagiography, as well as the character of the site itself, that the Israeli government deemed illegal. Though the government had banned the Kach Party from Israeli elections and outlawed the public

Figure 15. The Goldstein tomb and plaza in Kiryat Arba, 1996. This photograph was taken before the plaza was demolished. Photo by author.

display of Kach symbols or materials praising Goldstein, his gravesite continued to generate celebrations among the religious far right. This galvanized the left-wing and secular Meretz Party to sponsor a bill banning shrines to terrorists in general, while targeting the Goldstein grave in particular. In 1999, a long six years after the massacre, the military was dispatched to bulldoze the plaza, leaving the grave standing in the rubble. Yet the inscription on it was allowed to remain pending an Israeli High Court ruling over whether the state could dictate what was written on tombstones. Settlers turned the bulldozing into a media circus, with Goldstein's father tearfully draping himself over his son's grave. The demolition therefore left intact the very conditions that had generated the massacre to begin with, allowing Kach's marginal ideas to gain far greater acceptance.[31] In this manner, the Israeli government sustained the conditions for religious radicalism and settler violence while at the same time trying to limit overt support for one of its leading advocates. The government's actions then prompted a debate over the rights of free speech and religious freedom and whether ideological settler views did in fact incite actual violence. This debate, in other words, focused on whether the state had the authority to set limits on religious expression and whether settlers were entitled to shape religious spaces as they saw fit. While mirroring wider debates on religious freedom in Europe and the United States at the time, these failed to address the distinct conditions that had directly led to the massacre.[32] Moreover, Israeli liberals and the left wing dismissed Kiryat Arba settlers as religious fanatics, while ignoring the religious adaptations and government backing that had led to the massacre and enabled other incidents of violence to take place.

If violence is often justified on religious grounds, then the case of the Goldstein massacre raises a number of key questions: Can one separate out an ideological settler mode of life from the force and violence that have been used to sustain it? Moreover, do the ambiguous legality of the military occupation and its direct links to ideological settlement help shape Jewish radicalism and settler violence in the name of a divine mandate? I have tried to relate small-scale acts of settler harassment and encroachment to a broad spectrum of violence that constitutes the settler-occupation nexus, including structural violence and the militarization of settler Judaism. The Goldstein massacre as analyzed here, then, was not just a random act of violence, but rather the product of more systemic violence grounded in asymmetrical power relations and the evolving spatial arrangement of a sacred religious site.[33] Military force, spatial partition, and the shifting terms of Jewish worship supported Goldstein's ideological vision of exclusive control, which he attempted to realize through violence.

The Goldstein massacre also has to be seen in relation to a lived disdain for the other, as well as any limits set by law. These, too, had religious and spatial dimensions that articulated with a history of expansion, encroachment, and confiscation of Palestinian land that turned the Tomb of the Patriarchs into a crucible for religious violence. The apportioning of space inside this site to either Jews or Muslims in turn generated and confirmed ideas around social difference as fixed and opposed. Moreover, the impending sense of threat among settlers that contributed to the massacre did not just come out of an uncertain political period during Olso, but rather from an unsustainable vision of religious rule over Palestinians.[34] At the same time, multiple social and political divides were being generated within the settlement through new waves of immigration and conversion, and a number of these intra-Jewish differences also began to be perceived as threatening to settlers. It is to these internal differences, along with settler perceptions of the threat to religious authenticity that they presumably pose, that I now turn.

Chapter 6

Lost Tribes and the Quest for Origins

Because the ideological pull of settling does not readily attract new recruits as it once did (e.g., as during the height of the Gush Emunim movement), the push of economic incentives has often brought, or been used to bring, vulnerable immigrant settlers to peripheral and conflicted settlements in the Occupied Territories. Yet economically precarious immigrants moving into ideological settlements introduce social and cultural divides that threaten to dilute the existing sense of unity within the settlement. This chapter explores these intra-Jewish divides and the complicated ways in which groups presumed to be original Jews are used to override differences introduced by a diverse set of immigrant groups. Specifically, it shows how a religious allegiance to Hebron rather than to Israel proper is cultivated and required as a requisite of absorption into Kiryat Arba, even though most immigrants do not share religious beliefs or histories that readily link them to this settlement. A good number of new immigrant settlers do not even come from traditionally Jewish backgrounds. Therefore, their inclusion in an ideological settler community is based on a process of conversion, which is the product of a new missionary zeal. In this internal process of differentiation and integration, understandings of ethnoreligious difference are worked through on multiple and overlapping levels. These in turn are significant because they translate into both a demographic battle and a distinctly immigrant version of disdain for Palestinian difference, heightening rather than diminishing the wider conflict.

If previously I focused on settler reformulations and practices that enter into remaking existing Palestinian spaces as exclusively Jewish biblical sites, here I attend more directly to messianic "ingathering" or "return" as a newly conceived process of historic repair. I look at the ways in which some immigrants rather than others, generally recent converts rather than Jewish immigrant groups with a more extensively documented Jewish past, get directly interpolated into the core of an ideological settler project. In this process, a revaluation of place-based attachments and biblical origins widens the parameters of what counts as the Jewish diaspora, while concentrating a more diverse array of people and pasts into a small-scale settled locale. The irony is that observant Jewish settlers are often willing to create ties to immigrant groups that are not officially recognized as Jewish by existing government or religious authorities within Israel proper, in spite of their overall intolerance for other political and social forms of "difference." This chapter, then, argues that these processes of absorption and differentiation ultimately produce stringent albeit less traditionally recognized forms of Judaism that come together around forms of Palestinian exclusion. It seeks to answer the question: How does the internal diversity of new immigrants in settlements like Kiryat Arba get used in the remaking of Jewish tradition and the shaping of resolute attachments to biblical sites? And it ends with the idea that the quest for Jewish origins, whether through recuperated lost Jewish tribes or reestablished biblical sites of origin, is an ideological attempt to build forms of internal cohesion and coherence through a kind of nativist logic.

Before elaborating further, let me clarify some tacit assumptions around the issue of "difference" that factor into this ethnicizing process. First, when I talk about the "sorting" of differences among new immigrant groups within an ideological settlement, I refer to the values that settlers give to linguistic, cultural, social, and physiological differences and their categorization, rather than to their objective character. These attributions are imposed (but sometimes also self-ascribed) and then naturalized as "commonsense" views that define groups against one another (Barth 1998). Previously, I referred to similar sorts of ethnicizing processes that exist between settlers and Palestinians, constituted in moments of trampling on or storming through Palestinian space. The significance of focusing on this internal sorting process that seeks to produce the collective self, however, is that these forms of exclusion and incorporation, provide a counterpoint to directly violent encounters with Palestinians even as they reveal aspects of a colonial mind-set that are relevant to both cases.

Immigrant "absorption" in an ideological settler context like Kiryat Arba has become one of the most important ways of recruiting new residents, and has a devout inflection. The biblical rubric of the "ingathering of exiles" (*kibuts galuyot*) intersects with, but ultimately departs from, much of the Zionist focus on immigration because, in the religious settler case, the aim is to draw newcomers into the fabric of Orthodoxy alone. The integration of new immigrants becomes more directly enmeshed with making settler claims to biblical sites of origin and realizing a messianic vision. That is, this form of ingathering involves the active negation of a prior exilic period through attempts to fabricate resolute bonds and attachments to reclaimed biblical sites in lieu of national territory. It elevates the moment of an immigrant's "return" to a collective restoration of Jewish renewal. The irony, of course, is that living in an ideological settlement creates ruptures of its own for immigrants, replacing "exile" with new forms of alienation (Neuman 2002).

To explore these dynamics further, I focus on three key immigrant groups that arrived in Kiryat Arba during the mid-1990s and changed its character. These include members of the Kuki-Chin-Mizo ethnic group, or so-called Bnei Menashe, from the region of Mizoram and Manipur in northeastern India, the Beta Israel of Ethiopia, and finally the Russians, who were by far the largest immigrant group. Through an examination of these cases, I intend to explore how an internal ethnicizing process arbitrarily led to the selection of some cultural traits over others as inherently Jewish, and the valuing of model converts as the most "authentic" of all Jews. These intra-ethnic processes were not linked to any inherent cultural differences considered to define these immigrant groups prior to their arrival but rather were shaped by distinct forms of recruitment and resettlement. The "lost tribe" of the Bnei Menashe, for example, arrived through a private recruitment initiative. They were "discovered," financed, and brought to Kiryat Arba as tourists by the far-right rabbi Eliyahu Avichail, who recuperated them as a single "tribe," converting them to Judaism only after they had arrived. For this reason, they were beholden to him and favorably disposed to the ideological settler aims he championed. In contrast, the Beta Israel Ethiopian Jews were first recognized as a "lost tribe" by the Israeli government, airlifted out of Sudanese refugee camps, and then directly placed in Kiryat Arba's immigrant absorption center against their will. In spite of the fact that the Bnei Menashe and Ethiopians were both religiously observant "lost tribes," settlers viewed them as opposites, respectively revered and disdained for their attitudes toward settlement.

Both of these so-called lost tribes in turn were seen as counteracting the "scourge" of secularism that Russians introduced into Kiryat Arba. Arriving during the massive wave of immigration to Israel during the collapse of the Soviet Union, these new Russian immigrants were classified as neither lost nor found "tribes." Rather, from a settler perspective, they were merely inauthentic Jews who opportunistically made use of government-issued housing vouchers to remain in Kiryat Arba out of economic need rather than any deep ideological commitment. The question that emerges, then, is twofold: First, how did an insular ideological settler community use the idea of lost tribes to sideline differences deemed more threatening to the boundaries of the collective self? Second, how was the evolving internal diversity within the settlement used to further marginalize a Palestinian population and cut off threatening forms of difference?

The Making of the Bnei Menashe

The idea of bringing back members of the lost tribes who, according to the Bible, descended from the inhabitants of the Kingdom of Israel conquered and expelled by Assyrian invaders in 722 BCE, draws on the trope of reversal and remaking—casting contemporary Jewish settler expansion into Palestinian areas as a restoration of original biblical ties to tribes who once constituted the Jewish people.[1] Though biblical in origin, the "lost tribe" is also a recurring trope that has captured the Western (and Jewish) imagination for centuries (Kirsch 1997; Ben-Dor Benite 2009; cf. Cooper 2006). More generically, it has referred to any "uncorrupted" indigenous group that has had limited contact with civilization, particularly in European and American colonialism as well as Christian missionary exploits (cf. Ben-Dor Benite 2009). In each of these cases, Jewish origins have been attributed to indigenous populations as a way of linking all social and cultural differences to a singular biblical origin—the myth of monogenesis—and championing the idea that time and history necessarily repeat themselves (Kirsch 1997).[2] Westward-bound settlers in the American frontier, then, reputedly understood their homesteading efforts as similar to the Israelite return to the Holy Land, rather than outright conquest (cf. Koffman 2011). These homesteaders framed their relationship to American Indians as one to lost Jews who were not so much being conquered as overtaken by Christianity according to a well-established historical trajectory (ibid.). Within the contemporary ideological settler context, we similarly find both of these strands, biblical and colonial, operating in tandem. As in the case of the American frontier, the takeover of Palestinian land is being masked, though in this case, immigrants from distant places occupy the role of both settlers and the original natives of the land. The complexity of this shift means that in ideological settlement, "lost tribes" can be used to avoid facing the theological dilemmas raised by conquest, which is explicitly prohibited by Jewish law, but for settlers it also means grappling with the outer limits of Jewishness in an insular community.

To explore this process of conquest and reinvention further, let us consider how the Zo, also referred to as Shinlung or the Kuki-Chin-Mizo, of Mizoram and Manipur first became the Bnei Menashe.[3] As others have pointed out, a "lost tribe" does not simply exist to be suddenly found (Kirsch 1997). Rather it has to be actively produced through a conjuncture of different social and cultural frameworks. The particular kinds of engagements that brought this lost tribe into being in the context of Kiryat Arba reflect important aspects of a settler process of erasure linked to religious conversion. Today, the Bnei Menashe in Israel (and the Occupied Territories) total almost two thousand people, brought together from a variety of tribal groups living in the border states of Manipur and Mizoram.[4] They remain heavily concentrated in Kiryat Arba, though they were also settled in Gaza as well as other ideological West Bank settlements (Krieger 2006). According to scholars, they were animists who were later converted to Christianity by various missionary groups in the nineteenth century (Weil 2003; Egorova 2015). Those who decided to "return" to Judaism reputedly looked to one Manasia as their forefather, and this was interpreted as being a corruption of the Israelite tribal group Manasseh (which became Menashe) by an unorthodox Modern Orthodox rabbi with settler ties.[5]

Rav Eliyahu Avichail first set out in search of lost tribes in the East and found them in India (Avichail 1998). Within the ethnic Kuki-Chin-Mizo tribal groups, he found a splinter group who claimed to be Jewish, apparently in ways that were highly syncretic and mixed with Christian elements. Their "apprehensibility" as Jewish thus required his acts of translation.[6] No doubt this splinter group had an interest in being discovered by this rabbi who had the financial resources to relocate them to Israel. Mizoram and Manipur are areas marred by intertribal and ethnic conflict, as well as separatist movements, so the connection to Avichail would have been seen as a way of gaining access to a better quality of life.[7] It may also have been that the Bnei Menashe's willingness to identify as Jewish came out of a wider anticolonial struggle, turning elements of an adopted Christianity against their former missionary proponents of conversion (cf. Kirsch 1997).

Whatever the Bnei Menashe's underlying and varied motives, it is clear that their claims to having Jewish origins were first made apprehensible to other ideological settlers by Rav Avichail himself. In 1975, he established Amishav (My Nation Returns), an organization that allegedly had the blessing of the rabbi Zvi Yehuda Kook at Yeshivat Merkaz ha-Rav Kook (Halkin 2002:8). Avichail claimed that Kook advised him to actively seek out these and other Jewish communities, which had lost touch with their Jewish roots, in order to return them to Israel. Funding for this venture was then secured from the right-wing U.S. philanthropist Irving Moskowitz (Moskowitz 1998:280). Initially, Avichail was seen as a crackpot embarking on quixotic adventures to remote areas in search of nominal Jews. Yet over time, he was able to gain popular acceptance by recruiting people who seemed to fit well within imagined settler conceptions of original Jews. First, in practical terms, he conferred a biblical identity on the Bnei Menashe by naming the group, and he then facilitated their immigration by bypassing the laws and regulations as set out by the Israeli Ministry of Absorption. At a later stage, he actively created settler-oriented venues for the Bnei Menashe's religious education and conversion, first in Israel, then in India and Nepal (Eichner 2010).

Avichail pressed for the Bnei Menashe to be housed in Kiryat Arba, and the settlement provided them with the opportunity to study, convert to Judaism, and gain Israeli citizenship. It was as if bringing the Bnei Menashe to Kiryat Arba made Jewish claims more apprehensible through the medium of visible ethnic difference and provided a model immigrant group to offset the settlement's problems accommodating other immigrants who were less than enthusiastic about its ideological aims. Just as the remaking of sacred Jewish places required material inscriptions of the past to make them appear real, so too did the shaping of a Jewish settler peoplehood require palpable evidence of origins. The Bnei Menashe filled a "savage slot" by literally bringing back to consciousness practices of Jewish sacrifice and levirate marriage among others, and, in this manner, their presence paradoxically validated the ideological significance of reinstating "situated" Temple-centered Judaism. In contrast, the Ethiopians, or the Beta Israel, who were brought to the government absorption center in Kiryat Arba did not fulfill the "lost tribe" role in the same way. Not only did their ways of observing Judaism directly unsettle the idea of a Hebron-oriented point of reference, but the Ethiopians' actual dissatisfaction with life in Kiryat Arba seemed to raise questions about its Jewish authenticity.[8]

Rav Avichail's Visions

During my early fieldwork, I was repeatedly advised by longtime residents of Kiryat Arba to seek out the Bnei Menashe because of my anthropological training.[9] So, reluctantly and with some skepticism, I did meet with a few members of this group, which numbered sixty or so people at the time. To my great surprise, I found that far from being a coherent group, the members of the Bnei Menashe did not even share the same language. Rather, they spoke various Tibeto-Burman languages.[10] For this reason, they struggled to express themselves not only to outsiders but to one another, using broken Hebrew as their lingua franca. They had moved into the least desirable and cheapest housing in the settlement—rusting caravans, trailers, or temporary housing, *karavanim*—bordering Palestinian lands that settlers claimed belonged to Kiryat Arba but had not been permanently incorporated within its boundaries. Adding to these material and linguistic hardships were very difficult working conditions. The Bnei Menashe were employed, but only in menial labor. After meeting them briefly, I traveled to Jerusalem to talk directly with Rav Avichail, the rabbi responsible for bringing them to Kiryat Arba. He was an elderly but eccentric character who appeared to embody the imagined sense of Jewish authenticity he had long sought out in the wider world. He wore a large knitted skullcap and had a long white beard, and the walls of his modest apartment were lined with religious books, including the many volumes of the Babylonian Talmud, and biblical commentaries. I learned that he had received his ordination together with other rabbis of the Gush Emunim (Bloc of the Faithful) from the well-known Yeshivat Merkaz ha-Rav Kook in Jerusalem. By his own account, the "lost tribe" idea was his contribution to a religious social movement that seemed to be losing relevance and whose youthful energy had been overtaken by a routinized and bureaucratic mode of building settlements. Rav Avichail, in contrast, wanted to do something new and was therefore highly energized in his pursuit of novelty. In conversation, he confidently linked his quest for lost tribes to an array of biblical and rabbinic references, as well as other sources.

Rav Avichail's quest constructed truth in ways more complex than the simple literalism of other fundamentalisms, where the Bible is the word of God and the interpretive range of a text is vastly curtailed. In his case, biblical truth became apparent through the interpretive act of piecing together evidence from a variety of textual passages as well as clues beyond the text itself.

The textual references to the lost tribes "fit" with the Tanakh (Hebrew Bible), asserted Rav Avichail, and while he acknowledged that a degree of conjecture was involved, his understanding of truth meant that the sources all confirmed what he deemed to be reality. For instance, the passage in Isaiah he repeatedly quoted appeared in a poetic exhortation offering hope to a subservient, abhorred, and exiled nation. It mentioned protection from wind and sun, as well as guidance to springs of water and passes over mountains: "Look! These are coming from afar, from north and the west. And these from the land of Sinim" (Isaiah 49:12). For Rav Avichail, this passage confirmed that the Bnei Menashe were indeed originally from China, as they claimed, and that their Asian features had been noted thousands of years ago in the Bible. He added to this source the Shinlung's own ancient oral tradition that they had lived in the caves of China. Though the Bible referred to the tribe of Manasseh in one passage and then, in another, to the return of people from Sin (interpreted as modern-day China), it is Avichail alone who created a seamless link between biblical Sin, the lost tribe of Manasseh, and the Kuki-Chin-Mizo group he sponsored, who, he claimed, were Jews of the lost tribe returning to Israel from China by way of northeastern India.

Whenever Rav Avichail mentioned the spread of Jewish influences eastward, he also mentioned population size. It was as if recovering Jewish roots was also a matter of augmenting the scope and power of the Jewish population. He mentioned, for instance, that the Bnei Menashe were not the biggest group of lost tribes to be sought out, but that the Pathans (the ethnic group composing the Taliban ironically) of Afghanistan and Pakistan were much bigger. He estimated that there were thirty-five million people "who live as gentiles with signs of Judaism" in the East. Other rabbis in this settler fundamentalist camp would later contend more directly that large demographic numbers were key to the prosperity of the Jewish people. Rav Eliahu Birnbaum, a religious right rabbi linked to Shavei Israel, a successor organization to Amishav, which sought out lost Jews in Latin America, noted, quoting Genesis, that when God says "be fruitful and multiply," the intent was to have "descendants as vast as the sky is filled with stars." In his estimation, it was not sufficient to be a small and good community, but rather it was necessary to have a large and good community. His rejection of Judaism's status as a minority religion reveals new rationales for proselytizing and conversion, which is a significant change in Judaism because of its traditional stance of discouraging conversion (cf. Livneh 2002). It has often been pointed out, for instance, that in the biblical book of Ruth, Naomi discourages Ruth's conversion three times. To bypass this injunction and as a precaution to ensure that Jewish law had not been corrupted, conversion practices advocated by these rabbis and other supporters were framed not as conversion per se but as an attempt to get those who were originally Jewish to return to Judaism (cf. Bernis 2009). Conversions to eliminate doubt (*giyur le-ḥumrah*), in other words, were exponentially multiplying, and they essentially framed the convert as already Jewish but falling short of fulfilling the letter of Jewish law—merely fixing errors from the misconceptions that came about because of contact with impure contexts in the Diaspora.

What exactly were the signs of Judaism that Avichail found, and how did he know that they were not the sort of beliefs and practices shared across a variety of cultures and religions? When I asked him this, he responded that he had asked the chief rabbi of Israel a similar question: "It is written in the Tanakh that there will be signs of Jewishness—how should we know if these signs suffice?" He alleged that the chief rabbi said that "the more signs there are, the more you are compelled to help" and interpreted this to mean that any ambiguities would be resolved by a greater quantity of signs. Indeed, Rav Avichail went on to provide a list of extensive criteria that he used for determining Jewish origins. In the case of the Bnei Menashe, the criteria included circumcision, sacrifice, levirate marriage, the banishment of lepers, cities of refuge (*'arei miklaṭ*), the purification of menstruating women, the use of striped religious garments, and the content of a number of old prayers, whose actual histories were now lost to eternity. These original signs not surprisingly featured situated temple practices of pre-exilic Judaism that could be recuperated for a settler's resolute place-based attachments.

Irrevocably Jewish Signs

For Rav Avichail, the most important aspect of any Jewish sign was its authenticity. He felt that the combination of sacrifice, purification, and circumcision was distinctly Jewish and not shared by other cultures with non-Jewish origins. "With the Bnei Menashe, the Cohen [priest] comes on the eighth day, and they take two sharp stones, and put them in the fire to sanitize them, and after that they circumcise. No Jew does this now, but in the Tanakh there are things like this." It was telling, he thought, that circumcision among the Bnei Menashe occurred on the eighth day. He mentioned, as an aside, that Muslims also circumcise, but in a different manner and not within the same time frame, distancing the Bnei Menashe from Muslim practices and, by this opposition, distinguishing them as "original" Jews. Next he talked about biblical cities of refuge, which took in those guilty of manslaughter, noting: "In the Tanakh, it is written that if a man kills someone unintentionally, he should go to a city of refuge. The Bnei Menashe, don't have such a city, but they do have the house of the chief. So if someone kills unintentionally, he flees to the chief's house. While in the Tanakh, he takes refuge in the city for the entire life of a Cohen, among the Bnei Menashe, he serves as a slave in the chief's house." In other words, Avichail found slavery to be commensurate with the biblical mention of taking refuge in a city granting protection to refugees. Later I learned that the *bawi* system of indentured servitude to which Avichail was referring was a custom that no longer existed. In short, the Bnei Menashe were never a distinct group, and servitude had been a practice long abolished. It had been a form of flexible indentured servitude misinterpreted by a missionary presence as meaning permanent slavery that had been outlawed by 1915 (Zorema 2007). In the case of leprosy, another of the telling signs, Avichail claimed that the Bnei Menashe sent lepers outside the boundaries of the town, and while "no Jew in the world does this today," it too was evidence of Jewish origins: "If lepers go to a chief and he sees they are cured, he will tell them go to the river to submerge. In the Bible, a sacrifice is made with two birds, but among the Bnei Menashe they only use one. The chief will throw the feathers of the sacrifice into the wind, just like in the Torah, but not exactly. History has made changes but there is no nation in the world that does something like this."

Continuing on the subject of sacrifice, Avichail pointed out that the Bnei Menashe carry out a type of sacrifice in which they are forbidden to break the animal's bones, and he mentioned that this corresponded with injunctions in the Tanakh around the Passover sacrifice. "They call the bone 'halenkel' in their language. I don't know what 'halenkel' means, but they come here every Thursday and when I ask them about it, they gesture 'Passover.' And they have a very old song about Passover. When the priest makes a sacrifice, he says 'yud, kaf, vav, kaf' [spelling one of the names of God]. They only say 'yud kaf,' but if I am a priest of the Bnei Menashe, it is forbidden to say this word, unless one makes a sacrifice." His emphasis on sacrifice preserved in the face of dispersal runs counter to the conventional idea that Judaism should no longer focus on sacrificial offerings but on the study of scripture.

Rav Avichail looked to the Bnei Menashe as a way of recuperating many of Judaism's missing elements from the First Temple period rather than the

Second Temple period of other fundamentalists. He failed to consider that the contemporary Christian background of the Kuki-Chin-Mizo group and their familiarity with the Old Testament potentially provided them with material for the Jewish resemblances he observed. When I politely suggested that there was a certain vagueness to his findings, he insisted that the real problem was the overall lack of Jewish education in the United States, and that it was my own ignorance of Judaism that prevented me from seeing that it was absolutely clear that all the signs he found were undoubtedly Jewish. Then he chanted what he claimed was an ancient Bnei Menashe liturgy that included Hebrew place-names as irrefutable proof of Jewish origins:

ʿAneni, ʿaneni, Hoi Yah
ʿAneni, ʿaneni, ha-shoḥen be-Har ha-Moryah
ʿAneni, ʿaneni, ha-shoḥen be-Yam Suf (ha-yaduʿa)
ʿAneni, ʿaneni, ha-shoḥen be-Har Sinai

(Answer me, answer me, God
Answer me, answer me, he who dwells in Mount Moriah
Answer me, answer me, he who dwells in the Red Sea
Answer me, answer me, he who dwells in Mount Sinai)

Avichail claimed that the Bnei Menashe also had songs that mentioned their dispersal and exile in Afghanistan, suggesting evidence of a strong relationship to the Pathans, whom he also hoped would eventually return to Israel. His claims mark a significant shift in who counts as Jewish among observant Jews. Just as physical spaces are remade in relation to an imagined biblical topography, so too are the parameters of Jewish peoplehood being refashioned. Yet at the same time, these practices paradoxically seem to run counter to more puritanical strands of a settler sensibility. For instance, some settlers hold the view that the experience of exile from the land and intimate contact with non-Jews permanently corrupted Judaism. Rather than preserving origins or celebrating the diversity of the Jewish experience, they hold that those living in the diaspora led Jews to heresy and sacrilege and that it has introduced alien ideas into Jewish practice, which are the root of evil today. A fragment of this sort written in English that I took from a pile of scrap paper in Kiryat Arba read: "Unless we halt the penetration of foreign culture, then all the false distorted alien thought and dreadful hodgepodge which has infiltrated into our midst will remain in the Torah." This was evidently part of a larger manuscript, but it aptly expressed commonly held beliefs on influences to Judaism deemed alien. The xenophobic side of ideological settler culture held that foreign contact was mainly responsible for distorting interpretations of the Torah and that settlers were attempting to purge these extraneous influences. How then does one reconcile this anxiety around foreign influence and difference with the contemporary quest to recuperate lost tribes?

The Return of Tzvi Khaute

Bnei Menashe immigrants overcome these xenophobic anxieties by emphasizing their inherited Jewish origins, even when they convert. Tzvi Khaute, a spokesperson for the Bnei Menashe living in Kiryat Arba, asserted that his turn to Judaism was only a religious matter and that he did not convert for any financial gain or out of political motivations. "We don't mind the material difficulties of being new immigrants, even working as sweepers," he said. "A person should always be connected to *ha-Shem* [God]. We are not going to die of starvation; rich or poor, it all comes from *ha-Shem*. To have *emunah* [faith] is the most important thing." He also emphasized that he saw no connection between the social conflicts in Manipur, where he had lived, and his embrace of Judaism. Rather, he mentioned, he had worked in one of the highest bureaucratic positions of the Indian Initiative Service, which he thought provided good opportunities for the advancement of disadvantaged populations. Yet he nevertheless had decided to give up these upwardly mobile aspirations so that he could immigrate to Israel.

In his view, the main struggle of the Bnei Menashe was one of recognition. This struggle had taken many years, and Avraham Poraz, Israel's minister of the Interior (2003–4), had stepped in and arbitrarily halted their immigration. Then, he recounted, a rabbi sympathetic to the plight of the Bnei Menashe (Avichail) had approached Rav Shlomo Amar, the chief rabbi of the Sephardi community in Israel, and requested that he recognize the community as Jewish. With the help of other supporters, the Bnei Menashe had then set up a yeshiva in Jerusalem. Rabbi Shlomo Amar had agreed to send three religious judges (*dayanim*) to Manipur, he asserted, in order to determine whether the Bnei Menashe's claim to Judaism was valid. These *dayanim*, according to Khaute, interviewed people about their level of observance, wrote a report, and submitted it to Amar, and on the basis of it, he then decided to fully recognize them as Jews. For the Bnei Menashe, Khaute insisted, "it was a dream come true." Once recognized in 2005, Amar allowed the members of the Bnei Menashe in Israel to finance another religious court (*bet ha-din*) in Manipur that would continue to convert people there so that they would be Jewish prior to immigrating to Israel. Yet he stipulated that whoever did immigrate would have to undergo a subsequent conversion so as to erase all doubt (*safek dat*) about their Jewishness and to ensure that Jewish law had been followed properly.

If this struggle for recognition mainly focused on Israeli religious and government authorities, a second aspect of it involved acceptance by mainstream Israelis themselves. Khaute recalled a time that he had been shopping for groceries in a Jerusalem supermarket and was summoned by the manager of the store as a Thai worker: "Thailandi, Thailandi," he said the man had yelled disrespectfully to him as if he were just a "guest worker" and not a "true Jew." This annoyed Khaute to no end, and he regarded it as a deep-seated form of prejudice. He asserted, "You can't judge a Jew by their face alone. It all depends on belief, *emunah*. Most Israelis do not know we are not Thai workers." Thai workers who had lived in Israel for years and whose children spoke Hebrew as a first language, in his view, did not have anything in common with the Bnei Menashe because they had chosen their path for economic reasons. In contrast, for the Bnei Menashe, he emphasized, being Jewish was not a matter of choice, but rather of historical inheritance. Therefore, they were compelled to try to live out their true identity as Jews. Tzvi Khaute's final assertion was that the settlement of Kiryat Arba was where he felt most at home. He mentioned that when he came to Israel, he wanted to see which places would suit him most, traveling to Tel Aviv, Haifa, Jerusalem, Beer Sheva, and Ashdod. Finally, he said he realized that he felt most comfortable in Kiryat Arba and emphasized that every other member of the Bnei Menashe he knew was happy living in Kiryat Arba too. He felt a connection to the land, to his roots, and to his forefathers (buried in the Tomb of the Patriarchs), and his wife was satisfied, he asserted, even though she had to work to help support the family. He then stated that he would have served in the Israeli army had it not been for the fact that he had young children and no close relatives to look after them.

It is not surprising Khaute felt most at home in Kiryat Arba. This is where residents of the settlement first gathered donations to house him and others from the Bnei Menashe. He was then sponsored to go back to Manipur to bring back more converts. The community in which a new immigrant is resettled, its social networks, and the resources it uses to integrate families create connections and obligations, while also actively shaping ideological orientations and loyalties. Khaute's embrace of a settler form of Jewish observance, the material supports afforded him by settlers, and his social acceptance contribute to the "location" he calls home. His sense of place reveals how resolute bonds can be created out of imagined pasts—and how settler forms of Judaism revere some differences while erasing others. When I asked Khaute about living in a conflict zone, he gave a philosophical response: "All over the world," he said, "conflicts emerge as a result of an identity crisis. Groups fight for independence. They don't know who they are. The Nagas, for example, didn't know who they were, and many Kashmiris originally came from a Jewish background. Pathans too are Jewish." As for Palestinians, Khaute asserted, "a hundred or two hundred years back, they too were Jews but chose Islam out of duress. They didn't know their true identity, and the conflict has come about today because Palestinians are in the midst of an identity crisis." What was particularly disheartening was that Khaute echoed Avichail's ideas about different societies all having Jewish origins, and he did not seem to have another way of understanding how his own identity struggles related to the plight of Palestinians or, for that matter, how the Israel-Palestine conflict had historically developed. The Bnei Menashe had essentially escaped one ethnic conflict only to be interpolated into another as internally colonized immigrant settlers, and they in turn were helping to further the colonization process through an enthusiastic embrace of ideological settler views.

The Bureaucratic Making of Lost Tribes

If bringing a "lost tribe" to a settlement entails not only seeking out remote communities claiming to be Jewish but cultivating their devotion and loyalty, then the legal and bureaucratic structures that underwrite these quests play a significant role in shaping "tribes" as well. Previously, we examined how the gray zone of the occupation allowed facts on the ground to be retrospectively recognized as legal, enabling illegal settler acts to be carried out in the service of expanding control and taking over Palestinian land. In the case of immigration, too, a dynamic of testing the limits of the law initially paved the way for the early stages of the Bnei Menashe's immigration. The Bnei Menashe were paradoxically introduced to settler Judaism by a less than legal form of immigration at the outset. Their abbreviated conversion after the fact made use of a loophole, or paradoxical division of authority, that conferred on the religious establishment of Israel control over matters of personal status, creating a direct pathway to citizenship. In short, rabbinical authorities have the power to grant immigrant converts full citizenship rights, which then entitles them to a host of financial subsidies.[11] Yet, at the time, granting citizenship to the Bnei Menashe on the basis of "lost" Jewish origins and housing them in settlements on Palestinian land was a charged political matter (Curtius 1994). What motivated religious settlers to help, and what does it illustrate about the ideological underpinnings of settlement? In order to answer this, let me turn to the explanation by an elected municipal official of Kiryat Arba of the way she used resources at her disposal to resettle the Bnei Menashe.

As a religious right activist from a North African background herself, Bella Gonen was placed in charge of immigrant "absorption" (*keliṭah*) in the settlement. Her status as a municipal official of Kiryat Arba allowed her to work in both an official capacity as well as a nonofficial one, bypassing the constraints of Israeli immigration law. She did not see herself as actually breaking the law but rather as fulfilling the spirit of the law, while serving the greater good of the religious community and settler interests. She proudly asserted that while other members of the municipality of Kiryat Arba had debated whether or not to take in the Bnei Menashe because of the "considerable monetary output" it would require, she made a decision to help realize Rav Avichail's vision. This required bypassing the Ministry of Interior, the Jewish Agency's Ministry of Absorption, and initially even the Chief Rabbinate of Israel while gathering together local donations for resettling the Bnei Menashe. To host them, Gonen secured *karavanim* and furniture on her own: "refrigerators, beds, and basic gear so that families could begin to live here." She mentioned that she was able to oversee this process because of her extensive prior experience running the absorption center for Ethiopian immigrants.

In the Bnei Menashe case, however, there were some stark differences. With the Ethiopians, "everyone recognized them as Jews and they were under the auspices of the Absorption Ministry, and we got all the budgets, and *ulpan* [language school for immigrants], and everything else that was needed—health, social security, so you simply had to look after them." With the Bnei Menashe, on the other hand, more direct effort was needed, and it required influence: "Because I am an elected official of the municipality, people accept what I request of them—all the institutions in Kiryat Arba turn to the council when they want money from it, so when I went to request help from them, they couldn't refuse me." In her remarks, one sees evidence of official structures and illegal activities coalescing to form an integrated whole and the ways immigration resources and official policy were taken in a direction shaped by settler visions. That official and settler ideas around immigration were not entirely aligned became evident by the fact that government policy kept shifting with respect to the legality of Bnei Menashe's immigrant and citizenship status. Though the Bnei Menashe were not initially recognized as Jews or entitled to citizenship when they first came to Kiryat Arba, they subsequently did become legal for a ten-year period. They were then stopped from immigrating, and more recently the Ministry of Interior has taken preliminary steps to reverse that decision (cf. Lior 2015), exposing a rift around recognizing settler-sponsored religious immigrant groups for citizenship purposes.

Ethiopians: The Lost "Lost Tribe"

In some ways, the hosting of the Bnei Menashe in Kiryat Arba mirrored the resettling of the Ethiopians, most of whom had arrived several years earlier. Another nonwhite "lost tribe" who were discriminated against, Ethiopians were placed in an absorption center in order to be taught Hebrew, and educated not so much about Israeli life but about religious settler values and forms of observance. Yet as a rural agricultural people, they were not particularly enthusiastic about this reeducation, in part because their distinct way of observing Judaism was not being validated.[12] They looked to their own religious texts, some not found in the Bible, and their genealogies led back to Ethiopia rather than to Hebron as a site of Jewish origin. Moreover, their sense of being Jewish had been shaped by ongoing interactions with Christian and Muslim communities (Fenster 1998:182). When war broke out in Ethiopia, Ethiopian Jews fled to refugee camps along with others from different religious backgrounds, and they lived in these mixed communities until 1984, when, under threat of rebel forces, they were airlifted out of Sudan by Israeli jets (ibid.; T. L. Friedman 1985). Another mass airlift occurred in 1991, as the Ethiopian government was about to be toppled (Abbady 2015). Though Ethiopians had a far longer history of Jewish observance—believers date their beginnings to the fourth century—they were seen by settlers as opposite from the devout Bnei Menashe. In other words, while the Bnei Menashe were esteemed as exemplary Jews, the

Ethiopians were seen to be backward. These two groups nevertheless had a number of things in common: they originally came from war-torn and mainly rural areas, and their approach to Judaism was syncretic and deeply influenced by contact with other religions, especially Christianity. The settler assessments of these two groups, then, were arguably more a product of their "suitability" for furthering ideological aims than any social or cultural characteristics that actually defined them. While Kiryat Arba settlers became deeply invested in elevating, uplifting, and improving the educational and cultural status of Ethiopians, their efforts to bring them into the "present" often had the effect of turning them into permanent second-class citizens (E. Herzog 2007). Moreover, Ethiopians in Kiryat Arba were discouraged from adopting liberal and more democratic views that characterized secular sectors of Israeli society. Introducing them to a "religious modernity," then, essentially meant orienting Ethiopians toward a putatively authentic Jewish past that was entirely at odds with their own beliefs and history, while trying to bring them into the fold of practices aimed at taking over more Palestinian land in Hebron.[13]

Bella Gonen directed the absorption of these Ethiopian Jews in the early 1990s. When they were placed in Kiryat Arba's absorption center at the behest of the government, she focused on teaching Ethiopians skills she deemed necessary to lead a "modern" Jewish settler lifestyle. Much of her reeducation of Ethiopian immigrants focused on teaching them to use domestic appliances correctly. Gonen recalled introducing them to their housing: "I told them this is a house; this is like your *topel* [mud hut],[14] not to make an exact comparison between the two." She also recounted, patronizingly, the fear and wonder that Ethiopians had shown with what she presumed to be their first encounters with technology: "I opened the faucet and they were startled. I told them it's water, and an old man ran in back of the living room to see where the well was." The same, she asserted, happened with electricity: "I flipped the switch, and the light went on, and they were really scared. They were primitiveness personified." Gonen used well-worn tropes of fear and awe to describe this immigrant group. She also emphasized the "labor" that went into resettling Ethiopian immigrants: "I taught them how to light the stove, light a match. . . . We had hard labor [*'avodat perekh*] to teach them what a refrigerator is—they used to put clothes in the refrigerator and food in the closets. We taught them everything, from A to Z, toilets that you sit on and so forth. In the beginning, at the absorption center, they would go behind the bushes. We taught them why it's good to cook, and in pots and pans, and we also brought them washing machines that you put a token in."

Her narrative of backwardness says little about Ethiopians themselves and far more about negative settler attitudes toward difference and the ways Ethiopians clash with the domestic and gendered orders that presumably shape settlers' daily lives. Yet Gonen's anxiety and preoccupation with integrating Ethiopians into the settlement is instructive: for her, the Ethiopians seemed to be positioned at the outer limits of what it meant to be Jewish. This led her to talk extensively about how cut off Ethiopians had been from (European) contact and how they thought blackness was a Jewish norm: "Ethiopians say they are descended from King Solomon and the Queen of Sheba. And they were so disconnected from all Jewish communities for two thousand years that they believed they were the only Jews in the world, and that Jews were *only* black. And they believed that the land of Israel was a land gushing with milk and flowing with honey, where a person sat under his grapevine and fig tree—this was their idyllic [view]."

The Ethiopian past, as Gonen related it, was used to explain why Ethiopians were devastated by corruption in Israel and why they in turn also had trouble adjusting to life in an ideological settlement. Other "irrational" facets of Ethiopian belief that she referred to included their exceedingly strict ways of observing the Sabbath. The problem, she mentioned, was not only that young Ethiopians were becoming secular, but that the old generation observed Judaism in absurdly strict ways. For instance, Gonen gave the example of how one Ethiopian family, before relocating to Kiryat Arba, had chosen to observe the Sabbath rather than getting their son the medical help he desperately needed after he was bitten by a snake: "For the Ethiopians, mortal danger does not postpone the Sabbath. They kept the religion and tradition so much, they don't postpone. The father waited until the end of the Sabbath, and laying his boy down on a donkey, he walked five hours until he arrived at the closest clinic and by that time, the child had died." She conceded that Ethiopian Jews were religious, but not in ways that one should readily embrace. In contrast, her narrative suggested that settlers in Kiryat Arba set more rational limits on Jewish observance and knew when to break the Sabbath—perhaps an ironic view given the injuries and fatalities that settlers have incurred in the course of defending their vision of Hebron.

Much of Gonen's discussion then led to a final point, which was that the Ethiopians were a group in crisis, and that their encounter with modernity had shaken them to the core.[15] She had a simple explanation for this:

patriarchal authority, which she asserted had held Ethiopian families together in rural contexts, no longer held the same sway in a modern Jewish settlement. Parents, in her view, had no authority to prevent their children from embracing the worst aspects of Israeli culture. "There has been a very difficult crisis [*mashber*] occurring in the Ethiopian community, because young people rebelled; they don't listen to their elders, they threw off the kippa and stopped being religious [*zarḳu et ha-kipah ve-hifsiḳu li-heyot datiyim*]—[wearing] earrings and [counterculture] haircuts and all sorts of things. . . . They want to imitate Israeli society but in the most negative manner possible."

Her observations speak to the overall alienation of many immigrant Ethiopians. Yet Gonen's sense of the Ethiopian crisis overlooked the burden of their being scattered throughout various absorption centers in Israel, far from other family members (E. Herzog 2007). Ethiopians were subjected to ongoing bureaucratic scrutiny, which bred misconceptions that further led to their marginalization (ibid.). News coverage of the time mentions that most of the Ethiopians brought to Kiryat Arba felt trapped; nearly all of them wanted to leave, but they could not get government assistance to do so (Brinkley 1991). Settler tendencies in assessing intra-ethnic differences within ideological settlement, then, meant that practices that approached, imitated, or confirmed core settler values were embraced, while those that explicitly called living in Hebron or the process of settling into question were marked as alien or backward. Integrating the Ethiopians and the Bnei Menashe, both nonwhite communities, within the settlement did not ultimately translate into the greater accommodation of social or ethnic difference. On the contrary, it produced an internally diverse and yet exclusive form of settler Judaism.

The Russian Overflow

So far we have considered forms of internal differentiation and the varied positioning of two minority Jewish immigrant groups within ideological settlement. As a comparison, and way of thinking through other aspects of this intra-ethnic process, I turn to the case of Russian immigrants in Kiryat Arba. With respect to Russians, by far the largest immigrant group to have resettled there, a similar form of marking and (de)valuation seemed to be at play, except that their lack of piety and nonacculturation elicited a discourse around "commitment" rather than "origins." Settlers did not feel that

Russians could really be counted on to integrate and lead an observant life even when they opted to convert to Judaism. In this way, they were seen as both like and unlike the other two new immigrant groups.

After the collapse of the Soviet Union, a wave of Russian immigration to Israel totaling nearly a million people brought many secular Russians to live in Kiryat Arba for economic reasons alone. Equipped with vouchers and no longer required to live in absorption centers, Russians were nevertheless drawn to the settlement because of its highly subsidized housing.[16] Most saw their residence in Kiryat Arba as temporary, and they hoped to move on to a better place when they became more established. Russians were therefore not committed to leading religiously observant lives, eliciting negative views among the settlement's longtime residents.[17] For example, Devorah, a widowed woman and first-generation settler, leaned toward me in the midst of animated Russian conversations taking place in one common area and stated: "I am completely against the Rusifikatsyah [Russianization] of Kiryat Arba." Her remark, which deployed a Hebrew neologism, sounded shrill against the sound of Russian being spoken. Presumably she was not against the presence of the many elderly Russian grandmothers congregating on public benches so much as their unwillingness (or inability) to integrate by trying to speak Hebrew.

In fact, Kiryat Arba had become bilingual overnight, and all of its municipal notices and leaflets were suddenly two-sided, printed in both Hebrew and Russian. The older generation of Russian immigrants living in Kiryat Arba strengthened its right-wing tenor by supporting ethnic Russian political parties that developed within Israel to address their needs while participating in Russian-speaking social clubs. They read Russian newspapers and listened to radio in Russian, following news about both Israel and Russia in equal measure. Their ambiguous religious status further contributed to tensions and their separateness. While in Russia, they had been categorized as Jewish by virtue of having been born to a Jewish father but in these religious settler circles, they did not count as Jewish unless born to a Jewish mother. The same ambiguity around their religious identity held true within Israel proper, but in the devout atmosphere of the ideological settlement itself, the emphasis on leading an observant settler life, compounded the alienation. Even though the Israeli government, then, had granted Russians Israeli citizenship before their arrival, they were not well received in Kiryat Arba as Jews because they were deemed to be secular.

The divergence between Israeli immigration law and religious law posed significant quandaries for ideological settlers in ways that differed from that of the other two immigrant groups. The state had extended the "right of return" and citizenship to many Russians on nonreligious grounds: those who were considered Jews by virtue of their paternal line, or by intermarriage, or even those distantly related to a Jew were eligible to become citizens. The state's ostensible rationale was to recognize the "actual situation" of Jews in the former Soviet Union rather than impose stringent Jewish standards on them, allowing Russians to preserve the integrity of intermarried families. In Kiryat Arba, however, not only were these Russians not considered authentically Jewish, but the mix of non-Jews among them introduced an overtly Christian element into insular settler circles. Russian immigrants, then, reshaped the boundaries of intra-Jewish difference as did other immigrant groups, but they differed by introducing overtly Christian followers into the fold and weakening earlier settler claims to Hebron based on direct ties to Jewish origins alone. This mixed Russian settler presence in turn threatened to blur the ethnic boundary between Jewish "self" and Palestinian "other" since Palestinians too were composed of a Christian minority. Had the Russians been a small, powerless group, the Russian "threat" might have been overlooked. Yet, unlike the Bnei Menashe and Ethiopians, their large numbers and distinct social networks made it difficult to assume that Russians, whether Jewish or not, would be easily assimilated. There are no precise figures available on the number of non-Jewish Russians who came to live in settlements, but it has been estimated that up to 30 percent of all Russians from this period did not identify as Jews (Lustick 1999:419; Remennick 1998).[18]

Among devout settlers, conversion to Judaism was considered to be the key means of integrating this population into its distinct social fabric. Yet integrating Russians meant drastically reducing the time frame required for them to convert and lowering the bar with respect to the knowledge of Judaism it required. Russian non-Jewish women who did convert, a distinct minority, told me that it was very easy—there was no minimal time frame (most studied a few months as opposed to over a year) and only two letters from neighbors were required testifying to their sincere intention to live according to a Modern Orthodox way of life. No fluency in Hebrew was required because Russian translators were present at the ceremony, and the convert was simply asked to summarize a few basic tenets of Judaism listed at the front of their prayer books: belief in monotheism, the prohibition against idolatry, eating kosher (no pork), and keeping the Sabbath.

These expedited Russian conversions soon elicited controversy in settler circles. A telling discussion ensued in the now-defunct religious settler journal *Nekudah* under the title "Li-Deḥot o Le-Ḳarev?" (To Reject or Embrace?). The discussion, which features the opinion of several prominent rabbis, concludes that Russians should not be held accountable for intermarriages and non-Jewish offspring due to the conditions imposed on them during the Soviet era. Moreover, it suggested that the stance of the religious settler community toward them should be to embrace their difference and use novel forms of mass conversion. Under the subtitle "Is It a Commandment to Draw a Gentile Toward Judaism? Indeed, It Is a Commandment to Repel Him," Rabbi Joel Ben Nun, a leading figure in the moderate wing of religious settlers, addresses the quandary of reconciling the Jewish requisite of discouraging conversion with what was a rapidly growing Russian and non-Jewish presence in ideological settlements. He proposed turning to "mass conversions" accompanied by music, "an affair that ignites the exuberance of souls," while also pushing for greater restriction on allowing new waves of non-Jewish immigration (Stern 1997:39; cf. Lustick 1999). Each participant in this mass affair would then sign a form saying he or she "accepts the central matters of Judaism" while briefly stating "Shemaʿ Yiśraʾel," the pivotal Jewish pronouncement of adherence to monotheism (Stern 1997:39). In 2008, a controversy erupted over who had the authority to oversee the conversion of non-Jewish Russians, pitting Modern Orthodox rabbis aligned with settlers (led by Haim Drukman) against less lenient Haredi ultra-Orthodox rabbis. The Haredi rabbis who control the rabbinic courts and preside over conversions have tried to invalidate expedited conversions. They also have attempted to revoke the Jewish status of thousands of Russians who were expediently converted, prompting the secular High Court of Israel to weigh in on the matter (Ettinger 2012; Sterman 2016). In a rare intervention in religious affairs, the High Court voted to overturn the Haredi decision (Ettinger 2012). Moreover, liberal Israelis sided with the Modern Orthodox (settler) position as the more acceptable and accommodating approach in spite of its ramifications for expanding populations in ideological settlements.[19]

Part of integrating these Russians into settler society required not only getting them to embrace the "yoke" of the Torah but also helping them gain fluency in Hebrew so they could begin to read it in its original form. In a privately organized *ulpan*, or Hebrew language class, geared toward Russian housewives that I attended, class discussion often focused on the difficulties they were having adjusting to their new lives in an ideological settlement. The teacher, Simha Mizrachi, an activist *ulpan* teacher of Moroccan descent, saw her role as helping to integrate these Russian immigrants by providing them with new vocabulary, while inculcating prevailing Jewish settler values through her language lessons.

On the problem of "difference" within the settlement of Kiryat Arba, she led the following discussion.

> *Simha*: So maybe you can characterize the social composition [*mirkam ḥevrati*] of Kiryat Arba?
> *Tatiana*: Difficult to say. Because there are just so many different groups.
> *Simha* [*elaborating*]: From many different backgrounds?
> *Tatiana*: There are secular [*ḥilonim*] and religious [*datiyim*] people here. Within the secular group, there are also a number of other smaller groups.
> *Simha* [*correcting her*]: "Subgroups" [*tat-kevutsot*]—the secular group is composed of many subgroups.
> *Tatiana*: But to outsiders, we all seem alike. And there are also differences that exist between those who are born Israelis and new immigrants.
> *Simha* [*turning to another student*]: What characterizes the social fabric here in your opinion? Is it mixed and complicated? And despite this we live together?
> *Galina* [*thoroughly agreeing*]: *Because* of it we live together. If we were all the same, we'd be like cat and mouse!
> *Simha* [*rearticulating the point*]: The mixture enables us all to get along despite the differences, despite different backgrounds and despite different attitudes.

Clearly, Mizrachi tried to downplay the many internal tensions and divisions that threatened to destabilize the internal cohesiveness of the settlement, characterizing it as a veritable melting pot. Yet shortly after this pluralistic principle had been stated and elaborated, class discussion broke down around the controversial issue of settlers who were demanding government compensation to leave, giving rise to allegations that Russians were joining forces with other agitators who were "disloyal." Some Russians, in fact, did want to leave, and, like Ethiopians, they had requested government compensation to do so.[20] Their perceived lack of commitment was perhaps even more of a divisive matter than their unwillingness to lead a devout way of life. Mizrachi then proclaimed that anyone requesting government compensation was just an opportunist. She made clear that asking for compensation was a red line that should not be crossed and that the multicultural difference she had embraced moments earlier had its stark limits: "Now the question of compensation, from where did it spring up?" Then, introducing vocabulary, she continued, "'Spring,' *tsats*, 'to spring up,' *la-tsoots* . . . it came from the Sinai, Yamit, from the precedent of Yamit.[21] Those residents received compensation when Israel withdrew and so other people said, are we to be pushovers [*frayerim*] and not receive compensation?"

At the time, many thought that Kiryat Arba was slated to be dismantled and that immigrant settlers would willingly leave if there was adequate compensation. Mizrachi thought that anyone who wanted to leave should go, but only of their own volition. Accepting government compensation, in her view, meant ruining the ideological aim of allowing Jews to remain in Hebron. Mizrachi also believed that Russians should put their energies into advocating for a greater military presence to ensure greater safety for all and push for building permits to expand. Her strong objection to compensation reveals clear fault lines with immigrant settlers and also the palpable concern with being removed from Hebron.

The Diverging Paths of Russian Immigrants

Not many Russian immigrants managed to integrate fully into the social fabric of Kiryat Arba, but Inna and Yuri were exceptions. Like others, they had come from St. Petersburg and originally moved to Kiryat Arba for financial reasons. Inna would often say that they had never really *chosen* to move to Kiryat Arba, but once there, they remained and embraced its form of Jewish fundamentalism. Their religious turn meant greater acceptance within the settlement but more distance from other Russians in Israel, as well as secular Israelis in cities like Tel Aviv or Haifa. Their motivations for becoming observant settlers came out of a complex assortment of political, economic, and personal motives. Yuri, for instance, had divorced, and, leaving behind his former (Christian) family in Russia, he decided to start anew and build another life with his new wife, Inna. In Kiryat Arba, they purchased a two-bedroom apartment in an older tenement building and worked as self-employed artisans, using much of their living room as a workspace.

When I asked about their turn to a religious way of life, the couple noted that during the sixties many secular Russian intellectuals, as well as Jews, attended the Russian Orthodox Church as a form of social protest. Yuri said it was this countercultural aspect of religion that first attracted him. He mentioned that, as a kid growing up, he preferred Western over Soviet music and had been an avid fan of the Rolling Stones and the Beatles. His political and cultural sensibilities still retained this uncanny relationship between religion and protest, albeit now from the vantage of the far right. Embedded in his right-wing opinions were traces of a countercultural past that underscored in unexpected ways staunch anti-Arab positions. Being religious in a Soviet context, both stated, had been superficial. They illustrated their point with an example of one of their friends who suddenly decided to wear a cross but never went to church for confession or took communion. "How truly religious could he have been?" they asked rhetorically. In contrast, it was clear that they had fashioned their relationship to Judaism against what they understood to be these superficial gestures.

Within Kiryat Arba, long-time settlers considered this Russian couple to be outsiders despite their religious orientation because they were self-taught and Yuri had not been schooled in a hesder yeshiva. They also stood apart from secular-minded Russians by virtue of having taken a devout path. Moreover, Yuri voiced his frustration at not being able to find work in Tel Aviv, which he attributed to his religious lifestyle. Though the couple did not feel that they had been directly subjected to discrimination in Kiryat Arba, they did recount the case of an acquaintance who had considered himself Jewish in Russia because "Jewish" had been marked in all of his identity papers. They told me that after he immigrated to Israel he suddenly became a non-Jewish Russian because he had not been born of a Jewish mother. When he was told he would have to convert, he became so bitter that he decided to become Muslim. Their anecdote and its ironic twist seemed to indirectly speak to their own concerns about being accepted as authentic Jews.

Inna and Yuri participated in settler life not only through Jewish observance but by brandishing their far-right credentials. For instance, they spoke of how their daughter had been teargassed by the military for resisting the withdrawal from the Palestinian al Rajabi building that they and others had dubbed "Beit HaShalom," or the House of Peace (Weiss 2008). It was an event that received extensive media coverage outside of Kiryat Arba and had caused, according to Yuri, "total chaos." He recounted it as follows: A wealthy Palestinian owner who actually had wanted to go off to Jordan sold his broken-down property to settlers for far more than it was worth. The

PLO, he surmised, most likely then came and threatened him, and the owner was forced to take it up with Israel's High Court, clearly a "left-leaning" institution. Yuri emphasized that all the judges of the High Court should actually be wearing "Peace Now" (Israeli left-wing) buttons because they always ruled against settlers. In the case of Beit HaShalom, the High Court initially intervened and determined that settlers had purchased the property with major irregularities, and therefore those who occupied it had to be evicted (Glickman 2008).[22] The couple's attitudes toward the "biased" legal system and Palestinians expressed their allegiance to the religious far right in spite of their outsider status as devout Jews.

In contrast, most Russian immigrants had taken a comparatively secular path. Anya and her husband Vladimir, for example, exemplified this other route. They had been employed as engineers, and decided to leave Moscow during the collapse of the Soviet Union for economic reasons. They never had any intention of moving into a religiously observant community, much less an ideological settlement, but their financial situation led them to move to Kiryat Arba. Before immigrating, they had attempted to liquidate their assets and sell their coveted apartment overlooking the Moscow River for a large sum of money. As they were not the legal owners of the apartment in spite of having lived in it for decades, they decided to sell it on the black market. The sale was circumspectly carried out behind closed doors only a few days before their intended departure. Having carefully arranged the paperwork, they signed the apartment over for a large sum of cash, which they intended to use to buy new housing once they arrived in Israel. However, after the sale, they were immediately held up at gunpoint and ended up leaving without much money. Answering a knock at the door of their apartment and assuming that the buyer had forgotten something, they were robbed. They mentioned that they thought the buyer had prearranged the robbery to recover his payment. Because the sale was illegal, Anya and Vladimir could not go to the Russian authorities. When they arrived in Israel, then, they used their housing voucher to purchase an apartment in Kiryat Arba. After a few years, they bought another apartment in Maʿale Adumim, a so-called quality-of-life settlement located closer to Jerusalem. Therefore, they continued to live in a settler context, though they chose a mixed community that would be one more accepting of their secular orientation.

Due to housing shortages, however, their new apartment had not been built, and they continued to wait it out in Kiryat Arba. They did not find employment in engineering. Instead, Vladimir worked as a foreman, overseeing Palestinian workers on settlement construction sites, and Anya worked as a secretary in the radical right Jewish Quarter, located in Hebron's Old City, recording charitable Jewish donations. By virtue of their employment, the couple participated in furthering an ideological settler claim to Hebron even though they were very secular themselves. In order to fit in, Anya would dress for work as did other religious women, wearing a long skirt and covered head, but then she would change back to pants when she got home. While she often held skeptical opinions of those with whom she worked, she nevertheless conformed outwardly in ways that her employment required. Her overall perception was that the community of Kiryat Arba was belittling to Russians. She noted that people showed great surprise that she knew how to use a computer, and she complained that her neighbors thought Russians were as "primitive" as the Ethiopians. For her work recording donations, she was paid about eleven shekels an hour, the equivalent at that time of about three dollars, barely enough to purchase staples for the family. She also worried that her son Alex was being bullied by kids at school, and when I met Alex, sitting quietly with folded arms, he told me that he only had one more year left of living in Kiryat Arba to endure.

Conclusion

The play of Jewish difference that gets worked out among and in relation to new immigrant groups is instructive for a number of reasons: it reveals that a good deal of work goes into remaking, distinguishing, and emphasizing particular elements of Jewishness among those being recruited to inhabit putative sites of Jewish origin. In the case of the new immigrant groups discussed here, this entails creating a variety of social categories and assessments that come out of internal and external processes of inclusion and exclusion. So, for instance, while Russian immigrants often shared with the Bnei Menashe the need to convert in order to be recognized as Jews, only Russians remained suspect after converting. While the Bnei Menashe filled the "savage slot," by virtue of their devotional practices said to preserve origins, Russians needed to emphasize far-right anti-Arab attitudes as a way of fitting into ideological settler circles. The Ethiopians, by contrast, were ambivalently accepted because even though they were observant, they were not seen to be religious or committed in the appropriate way. Though designated as a "lost tribe," they were overshadowed by the Bnei Menashe, who alone took on the role of confirming settler *values* as model Jews and desired outsiders. The dispositions and tendencies within each group thus became exaggerated and congealed in this inter-Jewish process of differentiation and (mis)categorization. Ironically, the cultural, ethnic, and linguistic differences among settlers were often as great as, if not greater than, those that exist between (predominantly white) settlers and Palestinians. That is, for ideological settlers, the collective "self" continued to be seen as a coherent Jewish entity even as its actual social composition and religious practices became more diffuse, shifting to accommodate the variety of groups that had joined them in Hebron. The more pronounced these internal settler differences became, the more the emphasis on "origins" seemed necessary to provide a sense of cohesion. Origins, in other words, came to define the outer limits of Jewish peoplehood in the process of taking over and reinstating sites for Temple-based Judaism. Paradoxically, the more these claims to biblical Jewish sites and origins were realized in practice, the further removed ideological settlers became from the "authentic" modes of Jewish observance they claimed be reinstating.

Conclusion

Unsettling Settlers

The 2005 withdrawal of Israeli troops and settlers from Gaza remains one of most significant turning points in the history of post-1967 Israeli settlement. In the Gaza Strip, soldiers had been deployed throughout densely populated Palestinian areas (one of the most densely populated places on earth) while protecting the security of approximately seven thousand settlers who lived on big swaths of land near the sea, cultivating lettuce and other vegetables in industrial greenhouses. Compared to Hebron, many settlers who lived in Gaza went there out of mixed motives, including many "quality of life" and economically incentivized Mizrachi Jews (Jews of Middle Eastern descent). While Gaza had its share of ideological settlers, they had less of a religious claim to its territory because it had little recognized biblical significance. Nevertheless, the Gaza withdrawal mobilized thousands of young settlers from other religious far-right or ideological settlements that extended from Hebron to Elon Moreh. These second-generation settlers joined religious hard-liners who, in protest, refused to take any compensation to leave their homes (especially in Neve Dekalim and Kfar Darom), generating violent clashes and confrontations between settlers and soldiers, while eliciting deep social divides within Israel itself (Newman 2005).[1]

A week before the actual pullout, West Jerusalem was ablaze with neon orange ribbons fluttering in the wind like flames. This neon color, invoking the danger Gaza settlers felt, added to the already freighted atmosphere of protest. As I began interviewing people about the impending pullout in Kiryat Gat, a devout Jerusalem neighborhood, I made the error of referring to it as a *nesigah* or withdrawal. This was the term that had been used during the pullout of Israeli forces from Hebron a decade earlier. Every time I asked people what they thought of the *nesigah*, I got a rather confused look. Finally, my roommate pulled me aside and corrected me: "You should be asking about the *hitnatkut*, 'disengagement,' not withdrawal." Because they seemed to be synonymous terms, I asked her to explain the difference. *Nesigah*, she explained, "implies defeat—troops only withdraw when they lose in combat." Disengagement, on the other hand, was when neither side had been defeated and "military forces simply cease to be in combat." *Hitnatkut* expressed the idea that no battle had been lost. Yet it seemed that the full force of an internal battle had just begun.

In the days leading up to the *hitnatkut*, the settlers of Gaza began to pile into Jerusalem's central bus station. While still living in these settlements, they had been able to blitz the Israeli media with confident assertions of their eternal claim to the land and their brazen refusals to leave of their own accord. Yet most did leave at the final hour. When they actually got off the bus from a two-hour trip that placed them squarely within a different urban reality, carrying and cajoling their many restless kids, most of the Gaza settlers looked bedraggled, dazed, and confused. Then, overnight, accusing signs and banners of protest sprang up; many read: "I don't have a home, do you?" The implication was not only that the settlers of Gaza had been made homeless but that the state had betrayed them by forcing them out of their homes. This idea was reflected in the common settler dictum "A Jew doesn't expel another Jew"—disseminated, as if it were the eleventh commandment, on T-shirts, posters, bumper stickers, baseball caps, backpacks, CDs, and DVDs. The charge of betrayal came from the belief that the Israeli government had reneged on its promise (to settlers and others) to prevent Jewish expulsions that were characterized as similar to those of the Holocaust and Inquisition or by neighboring Arab countries. By drawing on these deeply engrained historical traumas and nationalist narratives, ideological settlers began to actively engage the most profound anxieties of the Israeli public.

Meanwhile, Gaza, home to nearly two million Palestinians, mostly descendants of 1948 refugees from areas that became Israel, was portrayed on Israeli television at the time as a place of looting and wild jubilation. Absent were the remarks of Palestinian refugees who might expose the paradoxes of settlers' claims to having become refugees. Others who had been displaced within Israeli borders, however, did comment on their status. A Palestinian citizen of Israel and resident of an "absentee" village inside the Green Line remarked that no municipal services had ever been extended to his village. While sympathetic to the plight of Gaza settlers losing their homes, he remarked, they at least had been well compensated and resettled while those in his village had been waiting for more than fifty-five years for recognition and

Figure 16. Women settlers and their children during a Jerusalem protest in 2005, before the Gaza pullout. Photo by author.

basic utilities.[2] Next, an elderly Bedouin leader living in the Negev appeared saying that while he welcomed the resettling of settlers in his desert neighborhood, his community deserved the same educational opportunities and resources that were being allocated to them.[3] As this complex tapestry of presence and absence was transmitted over the airwaves, it became possible for Israelis to collectively participate in the trauma of settler dislocation, while sidestepping the plight of over a million Palestinian refugees in Gaza alone as well as other marginalized communities within their borders.

If the lines of internal Israeli conflict had simply been contained within the neat rubric of religious versus secular, the removal of these settlements might have been destabilizing at best. Yet clearly the fault lines ran deeper, calling into question some of the basic assumptions and divides that lay behind the legitimacy of the Israeli national project itself. Some Israelis, including many liberal advocates of withdrawal, responded by expressing the view that removing settlers from Gaza would not serve to normalize relations with Palestinians. Their position was that it would simply begin a process of conceding more territory without solving the conflict. Removing settlements, they argued, would not change the fact that Palestinian refugees would still want to claim property within the Green Line, where Israelis now lived. Territorial concessions, in this view, would not turn back the clock and change the fundamental contradiction of a Jewish state built on land that had once been Palestine. There were others, however, who argued that remaining in Gaza was untenable and that leaving was the only decision that would ensure any chance of peace. To return to the settlers themselves, the withdrawal was not only a matter of relocating (which for the fifty-and-up age group meant starting over and reestablishing their lives at a late stage) but also of reevaluating ideological givens. Would settlers' injunctions to settle "bequeathed" land be invalidated, and would settling other Palestinian areas fall out of favor among their centrist Israeli supporters? The hard-liners among them vowed simply to return and start over, playing on religious themes of exile and redemption.

With the sudden influx of displaced settlers walking the streets of Jerusalem, the ideological battles affiliated with settlement began to shift inward. It led to a reexamination of not only the place of settler-oriented Judaism in Israeli public life but also of its wider influence. It was not as if this event exposed a singular rift between the Jewish state and ideological settlers, but it was certainly a most vivid illustration of the potential for the state-settler alliance to disintegrate into outright violent confrontation. Another comparable moment in Israel's history that mirrored this withdrawal was the 1982 dismantling of the settlement of Yamit in the Sinai. The evacuation of Yamit's settlers and the razing of their housing took place after Israel reached a peace agreement with Egypt. Yet once the Yamit residents evacuated the area, more ideological Gush Emunim protesters stepped in to take their place, attempting to fully dramatize the agony of relinquishing territory. The drama of the Gaza evacuation was similar, but increased tensions among a now entrenched settler sector led to violent confrontations that went far beyond this earlier protest. The number of troops deployed to Gaza, which according to some accounts left Israeli borders virtually unmanned, pitted Israel's entire military apparatus against the very settlers that the state had subsidized, defended, and supported for over thirty-five years. Moreover, a nearly five-thousand-strong settler presence and show of rancor at the protest (replete with acid tossed on soldiers) was now on display. Soldiers themselves were internally divided, and some whose own families lived in settlements had to decide whether to obey orders, which they did in the end.

When it comes to ideological settlements and outposts, withdrawals are likely if only for pragmatic reasons—many settlements are even seen as untenable by Israelis because of their proximity to Palestinian population centers, proving to be more costly than their political worth to the often right-leaning governments that sponsor them. Kiryat Arba and the Jewish Quarter in Hebron, for instance, had in the past been slated as candidates for dismantling in exchange for a peace agreement, though this is far less certain now. As the confrontation between settlers and soldiers in Gaza demonstrated, the positioning of many ideological settlers as both provisional allies of an existing state and religious radicals intent on undercutting state authority exposed glaring contradictions. In skirmishes over outposts such as Beit HaShalom and Amona, violent tactics by so-called "hilltop youth" also revealed both strong ties to existing settlements and anarchic tendencies that seek to move them in a new direction (Pelham 2009). These incompatibilities contribute to the increasing volatility of the state-settler equation, and as ideological settlers move up the ranks of the military, they are certain to play a significant role in shaping and determining the outcome of future domestic and international conflicts. While these past cases of confrontation show that settlers do not yet have the power to determine government policy outright, they nevertheless demonstrate the growing influence of this sector in social, military, and political arenas within and beyond Israel proper (cf. Pedazhur 2012; Pedahzur and McCarthy 2015).

Among the key things at stake in removing Hebron's ideological settlements are not only issues of social justice more broadly but also the long-term relation of these settlements to Jewish identity within Israeli society and beyond. Many ideological settlers and their supporters are convinced that they are operating within the framework of a long-standing Jewish past. In drawing out this narrative of continuity, and a historical right of worship directly in Muslim places (particularly present-day mosques), little attention has been given to the actual conditions and contexts that shape these settler forms of Jewish observance and how these differ from those of other historical Jewish communities. This ethnography, then, has grappled with various aspects of continuity and rupture, evaluating not only the ways Jewish tradition has been remade but its implications for equity, accommodation, and social justice with respect to a Palestinian population under occupation. One of the broader stakes of this analysis, then, is the challenge of dismantling both actual settlements and the ideological foundations that have enabled them to prosper.

There have already been strong objections to settler claims by Palestinians most directly impacted by them, as well as by liberal Israelis and moderates within its observant religious circles, among others. These groups have pointed to the threat to democracy, personal freedoms, and human rights that settlers pose, as well as the cycles of violence they elicit. Criticism within Israel has focused on the social and political costs of maintaining a military presence to maintain the safety of settlers, while, internationally, observers have pointed to settlements as key obstacles to peace. Some other critiques, however, have pointed a finger back at these critics and raised other concerns around religious expression and the rights of religious communities. For this reason, it has been worth engaging with the ideological dimensions of Judaism as it is being lived in Hebron—what its terms are, what social conditions animate it, the alliances it enables, how it bears on conceptions of temporality, and its material inscriptions—while also exploring the relations of power and social hierarchy introduced by the renewal of biblical sites, immigrant settler communities, and the reorganization of Palestinian space to accommodate them.

Ideology Revisited

This ethnography has shown that while ideology can at times be an elusive object of study, it is a critical lens for understanding a far-right religious vision. From the vantage of a settler context, beliefs are part of an ideological formation that encompasses conceptual, practical, and material dimensions. Within ideological settlements, Jewish fundamentalism entails a vision forged both under the auspices of the Israeli state and against the lives of Palestinian residents in occupied areas, but, as we have seen, it has also led to power struggles aimed at the authority of the state, liberal sectors of Israeli society, and different groups and factions within settlements themselves. This book has focused on the realm of micropractices actively shaping and shaped by the ideology of settlers in Kiryat Arba and, by extension, those of other radical religious settlements—investigating the lives, beliefs, and actions of those who reside in these armed communities out of religious conviction as well as economic need. It has examined the ways Jewish ideas and practices have been invested and enmeshed in an evolving biblical topography, one that situates Jewish origins and new forms of identity in sites that have equally long-standing Christian and Muslim histories woven into them. It has explored the way these sites of potential convergence have been separated and partitioned as a way of reintroducing a Jewish presence while expanding military control to surrounding Palestian areas. Settling for religious reasons has therefore required settlers to amplify Judaism's spatial character rather than its temporal, interpretive, and legal dimensions, emphasizing stark differences between rather than convergences among these other world religions. The finding that ideology is not just out there making itself felt in moments of crisis but rather is woven throughout multiple domains in the everyday speaks to its deeply entrenched character and is one of the key contributions of this ethnographic approach.

Approaching ideology in this way risks making it appear "terrifyingly normal" to borrow an Arendtian phrase. During the early part of my fieldwork, I sought out the advice of an established (and religiously observant) anthropologist teaching at Bar Ilan University on how best to document telling differences in an ideological settlement. Our meeting lasted perhaps half an hour at most, and he didn't have much to say about ideology but did offer the opinion that living in a place like Kiryat Arba conferred a good deal of power on very average people. Then, referring to Purim, the Jewish holiday that was approaching and that I expected would reveal insights through its role reversals and inversions, he added with a certain glibness that he doubted play was possible in a place like Kiryat Arba. He ended up being half right: I noted that while the kids were dressed in costume, they tended fall into two categories—boys outfitted as soldiers and girls dressed as Queen Esther but in wedding-like gowns as if they were also brides. These scripted and gendered roles indeed narrowed the parameters of play, while emphasizing a wicked plot to destroy all Jews.

The seasoned anthropologist left me with the following advice: "Look, people are going to show you how they paint rocks, and then you are going to write about it." He assumed, in other words, that a settler's distinct ideological inclinations would be quite evident to me at the outset. At the other end of this spectrum were scholars who questioned the efficacy of framing settlers as "fundamentalists," pointing instead to chauvinsitic tendencies in secular forms of categorization and misrepresentation. Yet in taking the position that these settlers were not all that different from liberal Israelis, they seem to have overlooked the real relationships that exist between religious convictions and actual day-to-day practices unfolding in assymetrical structures of power. Moreover, representing ideological settlers as identical to all other Israelis uncritically accepts far-right settler attitudes that normalize or minimize the direct violence of the everyday.

Fieldwork Dilemmas

The difficulty of deciding which aspects of daily life were most relevant for investigating an ideological formation in the field was compounded by engaging with and representing those with whom I had profound disagreements. At a personal level, I struggled with suspending judgment and setting limits. Nurit, for instance, a young mother in Kiryat Arba, tried to move me out of my observer role by making use of what she understood to be my leisure. She often asked me for help with childcare and domestic tasks, while giving me unsolicited advice about religious matters. Once, she asked me to cut her hair, a seemingly innocuous request that led to an internal dilemma. I certainly did not want to do it—we were not particularly close, and I was visitng her home to conduct research, or so I thought. However, she kept requesting this favor and insisting that no one would ever see the results because she always kept her hair covered in public as required for modesty. I finally gave in to her request, and she supplied me with a pair of small dull-bladed paper scissors. Before a cloudy mirror in the bathroom, I proceeded to cut her hair until she looked positively ragged. Then she took a kind of perverse pleasure in the whole matter, noting how ugly she looked. Was this her way of marking our differences or bridging an unspoken divide, I wondered? Was it saying something indirectly about her sexuality and the way it was channeled into arranged marriage and motherhood? There was something defiant about her joy in not being beautiful and making me an accomplice to an act of destruction. After I cut Nurit's hair, I disliked this close interaction, which led me to empathize with her in spite of our many profound disagreements over values and politics. This interaction also highlighted the degree to which a particular kind of aesthetics had shaped my own (gendered) secular sensibility. In short, my misgivings about something as simple as cutting hair, made me think about ideology as that which shapes even private decisions around marriage and child rearing, determining the character of intimate as well as public forms of interaction.

If setting limits on these private interactions elicited moments of awkwardness, observing public settler gatherings and contexts had other dilemmas. Settler violence has often been considered by scholars to be the result of ideological certainty facing a moment of crisis—that is, of deeply held beliefs colliding with historical events or peace processes that do not play out in their favor, requiring desperate attempts to salvage them. Overlooked in this standard explanation of violence as a response-to-crisis model are the small-scale but routinized practices that socialize settlers into indifference and detachment toward Palestinians. Many others have tried to make the case that violence is incidental to the kind of devout Judaism observed in Hebron and has nothing to do with religious sensibilities. This ethnography, however, has revealed a more direct relationship, using ethnographic examples to explore a spectrum of practices that contribute to violence in Hebron and other ideological settlements. Not only has it featured the many forms of violence that enter into the inscription of a permanent devout settler presence in Kiryat Arba, the Tomb of the Patriarchs, and the Old City of Hebron. It has also illustrated how the legal gray zones of the occupation and alliances with the soldiers enable direct acts of settler violence to go unpunished.

Furthermore, handguns given to settlers in order to shift the burden of security away from the military are often indiscriminately used with no oversight or repercussion. Guns are present in every sphere of settler life and beyond—in religious settler homes, on buses traveling to cities within the Green Line, and in unlikely places like Jerusalem's vegetarian restaurants and swim clubs, seemingly far beyond the boundaries of existing ideological settlements. The prevalence of guns also adds further dilemmas to a researcher's aims and obligations, foremost of which is to cause no harm. To the extent possible, I opted out of scenarios I thought might compromise the integrity of my research. For this reason, I decided to investigate settler violence by asking about the sense of threat that prefigures violence, as well as the many forms of complicity that arise in the aftermath of it.[4] I am aware, however, that others might have set different limits that would have precluded setting foot in a settlement at all. As a non-Israeli ethnographer and outsider, however, I decided to speak to settlers directly, as well as to a range of other accessible actors in the conflict, including soldiers, military authorities, peasant farmers, intellectuals, scholars, and, finally, Israeli and Palestinian human rights organizations documenting settler violence through testimonies. I crossed boundaries cautiously and during different periods of research in order to contextualize ideological settler views from the vantage of different angles. Yet it is also important to note that settler formations can exceed the actual space of an ideological settlement itself and meld with lives in everyday contexts within Israel as well.

One Saturday in West Jerusalem, for instance, I went to a swimming pool full of families trying to escape the heat and strictures of the Sabbath, when I bumped into members of a Kiryat Arba family I had interviewed the week before. I imagined myself on break from research and at a considerable distance from anything related to a settlement when they suddenly pulled up their lawn chairs and began chatting. I was first introduced to their American friend, a blonde woman with teased hair, who worked for a Christian Evangelical organization named (without irony) Bridges for Peace. This was an organization established to help devout Jews achieve their mission of returning to Israel in order to spur the Second Coming of Christ. The organization had funneled a good deal of money into a series of settlements, while also printing and distributing Bibles and preaching the gospel to anyone who would listen. This particular settler family, however, had come to the pool not to evangelize but to relax after dropping off their son for military service.

The father made a point of telling me that he thought I had been talking to the wrong kind of people and that I ought to seek out other residents living in the settlement (now referring to himself) who maintained friendly ties with their "Arab" neighbors. In terms of masking conflict and violence, this was one of the settler attitudes I encountered frequently—either statements about settlers having "Arab" friends or that many "Arabs" (never referring directly to "Palestinians") actually welcomed their presence. This man also proposed that I consider changing the focus of my research to "the ethics of a community under duress" in order to better reflect the fact that settlers were the ones under siege and were most often the victims rather than perpetrators of violence. Having said this, he then decided to join his family for a swim and asked if I would watch his bag of belongings, which, he added, almost as an afterthought, contained a gun. I quickly responded that I couldn't do it because that would mean going beyond being a researcher. He then abruptly pulled out the gun and walked off. Setting a limit here, as at other times, meant disrupting the conventions of a polite and, on the surface, congenial social interaction.

Ideological Settler Influence Beyond Hebron

Kiryat Arba and Hebron are local stages whose repercussions can be felt at the national and international level. Their insular space is a key arena for ideological production, but their impacts extend outward as well. This is not only because settlers move back and forth physically to Israeli cities and are increasingly part of Israel's national institutions but also because these settlements remain a hub of radicalization, recruiting and integrating different kinds of unlikely supporters and allies, including, as the example above illustrates, a network of U.S. evangelicals. Whether ideological settlements are expanded or dismantled, then, their influence has exceeded the role they initially played in creating a fragmented biblical and asymmetrical spatial order in Palestinian areas of the West Bank. Among a younger generation of settlers, a key dilemma in furthering their influence is whether to integrate into mainstream Israeli institutions or to continue to self-segregate in ideological enclaves (cf. Leon 2015; Lebel 2016). More recently, ideological settler practices have increasingly begun to shape the tenor of conflict within Jerusalem, and other areas within the Green Line as well.

Settler efforts to worship directly at the site of the Temple Mount/Haram al-Sharif in Jerusalem, in other words, have led to escalating tensions and random acts of violence between Jewish and Palestinian residents of the city. Among the so-called newer Temple Movements, there has been nothing short of "a dramatic increase" in the number and influence of organizations that are now focused on reinstating a Jewish presence directly in the area presumed to be the destroyed Temple (on the Temple Mount) and other currently Islamic sites (Ir Amim and Keshev 2013:75). These efforts run the gamut of "raising consciousness" around the significance of Temple worship in Judaism to actively aiming to rebuild the Temple (ibid.; Inbari 2009). The settler bid to pray at the exact location of the Temple, presumed to be the foundation stone of the Umayyad-era Islamic shrine known as the Dome of the Rock, rather than the standing Western Wall, attempts to remake the character of Jewish longing, erasing thousands of years of rabbinic Judaism. It also disregards the 1967 agreements between Israel and the Muslim waqf over the status quo of worship, as well as Jordanian guardianship over the holy sites. The Israeli government's response, in turn, has been vague with respect to limiting these settler tactics, variously monitoring and arresting extremists while actively supporting settler expansion and turning a blind eye toward these illegal activities. Ideological settlers have in turn used this ambiguity to push beyond existing regulations and legal limits, making the quest to rebuild the Temple the centerpiece of a seemingly more authentic Judaism that wields religious authority against the state.

As in Hebron, settler theological emphases have remade religious law and erased injunctions dating from medieval rabbinic Judaism to forbid worshippers (those who are deemed "impure") to tread on the destroyed Temple's sacred area (Inbari 2009). Meanwhile, this claim has occurred while religious-right settler activists take over and fortify houses in Jerusalem's mixed or exclusively Palestinian neighborhoods such as Abu Tor or Silwan, prompting critics in Israel to consider their acts forms of "Hebronization." These actions suggest the ethos of Jewish settler devotion as a mode of control and takeover is shifting from the Hebron periphery to Jerusalem and other shared religious sites. This illustrates Memmi's (1965:63) point that settler colonialism (in this case the occupation) presents a permanent danger for the home government. In other words, ideological settler practices devoted to undercutting state authority cannot readily be confined to militarily occupied settler areas alone but rather threaten to replace the already imperiled pluralistic character of Jerusalem's Old City. Further, attempts to solve potentially explosive conflicts over land and sacred sites with partition and separation alone (as in Hebron) preserve existing ideologies and power asymmetries. While the enforcement of separation through barriers and walls may therefore lead to the cessation of violence in the short run, it does not address any of its underlying causes.

Moreover, the new settler push to pray directly at the site of the Temple is occurring in conjunction with the proliferation of enclaves populated by the "hilltop youth," the radicalized children of more established ideological settlers, who squat with even fewer constraints on Palestinian property in the West Bank, far beyond the existing borders of ideological settlements (Pedahzur and Perliger 2011; Byman and Sachs 2012; Boudreau 2014; S. Friedman 2015). In examining these trends, most prefer not to look directly into the black box of religion, ignoring whether Jews are in fact obliged from a religious standpoint, as devout settlers claim, to reside in and pray directly at biblical sites that have long Christian and Muslim histories attached to them. Rather, these critical engagements tend to be outsourced to rabbinical authorities and observant communities instead of being taken up as complex challenges to social justice and human rights they represent in their own right. One finds, then, either a good deal of intolerance or reverence for all that falls under the sign of settler "devotion." In sum, there is a tendency to veer toward a nostalgic embrace of all that presents itself as "authentic" Judaism in spite of the darkening political landscape being ushered in under some of its guises.

Finally, there is the growing political influence of ideological settlers to contend with: what started out as a fringe group strategically playing on internal divides within Israeli society and successive Israeli governments to achieve its aims has morphed into a key constituency that effectively props up the mainstream right-wing Likud Party. In Israel's 2015 election, for instance, Prime Minister Netanyahu was more beholden to the far-right religious settler sector than ever before and reaped a victory in a tight race by directly appealing to deep-seated anti-Arab sentiment, alleging that he would be defeated by "Arabs" (meaning Palestinian citizens of Israel) voting "in droves." Others have pointed out that Rabbi Meir Kahane's xenophobic ideas, once banned from the Knesset, have also become more accepted in the mainstream (cf. Pedahzur and Perliger 2011). Several politicians espousing Kahane's ideas or explicitly identifying with his teachings have been elected to the Knesset, and Kahane is no longer the maligned figure behind the Goldstein massacre that he was once conceived to be (ibid.).

In sum, the realm of practice in ideological settlements such as Hebron has served as a social laboratory—testing limits, creating new realities, and instituting changes that have had far-reaching implications for not only remaking Jewish understandings of authenticity and Israeli nationalism but shifting the terms of the Israel-Palestine conflict from a dispute over land to one mainly expressed in an ethnoreligious register. The settler struggle to control more of the Tomb of the Patriarchs has subsequently been taken up by Prime Minister Benjamin Netanyahu in his 2010 declaration that the Tomb is national "Israeli heritage," even though Hebron lies outside of Israel's internationally recognized borders. Palestinians then countered his claim by nominating the Tomb as a UNESCO World Heritage site, invoking international law to protect it as an endangered site. At present, Israel's battle with UNESCO over the Tomb's "heritage" continues. Many contend that it is hard-core religious ideologuess who are the issue in Hebron and that they can be managed with better policing and law enforcement. Yet a different kind of debate is needed, one that addresses the range of Jewish settler expression and practice taking place under the banner of religious "authenticity." These matters are especially pressing when, as is the case presently, ideological settler sites that have been championed as sacred have led to exclusionary forms of control, prolonging the Israeli occupation and erasing both a historical and actual Palestinian presence.

Notes

Introduction

1. The terms "settlers" and "settlements," it should be noted, are not terms of self-identification among the residents of ideological settlements. They prefer to call themselves *anshe dat*, that is, simply "religious people." The Israel-Palestine conflict has spurred disagreements over what counts as a settlement. Most international observers consider Jewish housing tracts in East Jerusalem, an area annexed after 1967, to be settlements. Attempts to "Judaize" Palestinian-populated areas of the Galilee within pre-1967 Israeli borders are also considered by Palestinian citizens of Israel among others to be part of the settlement continuum. Most international observers and centrist or leftwing Israelis consider the settled Jewish areas of Hebron over the Green Line (the 1949 Armistice lines between Israel and its neighboring countries) to be settlements.

2. In this book, I show why ideological settlement constitutes a break with the past rather than a reinstantiation of it. When it comes to the Tomb of the Patriarchs in particular, the site has changed hands over the course of history, and many have insisted that it therefore cannot be considered a mosque alone. An Israeli anthropologist responding to an earlier conference paper of mine on the matter suggested that the site's pre-Islamic origins and sacredness for Jewish and Christian traditions rendered the identity of the present structure ambiguous. My response is as follows: First, I distinguish Hebron's textual invocation by three Abrahamic traditions from the specific physical site, and I take the identity of the standing edifice to consist not of its (incompletely documented) origins in antiquity but the way it has been continuously used over time for its longest duration. Because it was a mosque from the seventh century until the Israeli occupation of the West Bank in 1967, I privilege its identity as a mosque over any single period in history where political control shifted, power changed hands, and the edifice housed another religious community with its architectural features modified accordingly. No doubt this debate over which past counts most is socially constituted and variable, indexing who has the power to makes claims to a particular past and for what ends.

3. Basic modern Hebrew distinctions for settler and settlement include *mityashev* and the verb *hityashvut* (from the root *sv* as in the verb "to sit") and are used to refer to pre-1948 settler immigrants in Palestine, as well as towns within Israel proper, whereas *mitnaḥel* and *hitnaḥalut* (from the root "to inherit") is used specifically for a settlement established after 1967 over the Green Line in the Occupied Territories. *Hitnaḥalut* has a critical tone, which is why settlers refer to themselves as simply religious or *datiyim*, and their own communities with the unmarked term *yishuvim*. *Mitnaḥalim* are generally thought to be ideological settlers by Israelis. Indeed the division between quality-of-life and ideological settlers in the occupation can be more blurred than it initially appears. This is because ideological settlers, the vanguard minority of those living in the

West Bank, often come out of Israel's economic margins, while those in search of a better quality of life often take up religious rationales to justify their residential locations.

4. In what came to be known as the Allon Plan, conceived just after the June 1967 war and adopted with modifications, heavily populated Palestinian cities and towns in the West Bank would remain under Jordanian control via a corridor of land connecting Jordan River's East and West Banks. Israel would then rule over unpopulated land areas occupied in the wake of the war through a settler presence set up as an added barrier along the Jordanian border and other less populated areas in order to ensure its national security.

5. "Settlement" and "settler" frequently shift meaning depending on the perspective of the speaker. In the context of the Israel-Palestinian conflict, attitudes toward the category can range from some Palestinian perspectives that consider all Israelis to be "settlers" to many right-wing Israelis who maintain that there is no such thing as a "settler."

6. My consideration of religious place and its significance in settlement has been shaped by a central debate among the geographers David Harvey (1993) and Doreen Massey (1993). For Harvey, "place" is by definition reactionary. Bounding becomes a way of creating face-to-face communities that protect vested economic interests and guard against change. The conceit of bounding and fixing, then, in his view corresponds with inward-looking forms of consciousness. Doreen Massey, on the other hand, sees small-scale projects as potentially liberatory and tries to resuscitate the value of "place" in the wake of Harvey's negative characterizations by pointing out that places can open out onto a broad array of social networks and be sites (Massey 1994; Cresswell 1996, 2004). She embraces "place" in relation to its potential for progressive politics. While this ethnography is indebted to the debate, it takes a different position. Like Massey I see the analytical value of featuring "place" in ideological settlement, which creates exclusive enclaves, but, drawing on the work of Harvey, I treat settler practices of bounding as reactionary. Nevertheless, settling has ushered in far-right social and political transformations that make it a more dynamic medium than Harvey concedes.

7. For a classic discussion of reiterative time linked to "origins" in the context of Northern Ireland, see Feldman (1991).

8. What is perhaps distinctive about these resolute attachments is the counterintuitive sense of *not belonging* that often underwrites them. Albert Memmi (1965) in his classic essay on (settler) colonialism talks about how colonizers belong neither to the places they have colonized nor to the homeland from which they have emigrated. They literally have an in-between status in both of these locations. An ideological settler's belonging, too, is forged in relation to a distinct set of religious, political, and geographical orientations that most secular Israelis do not recognize as their own. This dislocation reinforces an ongoing quest to create permanent attachments to sites settlers claim as theirs alone on religious grounds.

9. Feige (2001) considered the place-based aspect of ideological settling to be an extension of Zionist practice. While I agree in general, characterizing it as Zionist alone greatly underestimates the devotional dimension of this form of place making and the subjective religious investments in biblical sites that underwrite it.

10. Smith (1992) argued that there were two paradigms of place operating in religious traditions: places that mark the site of a past event and therefore cannot be moved, and places that were significant as seats of power or religious authority and were transportable in principle. Judaism with its orientation around the Temple falls into the second category, and, for this reason, it morphed into a mode of observance that was replicable in any location.

11. Asad is working within a post-structuralist Foucauldian paradigm, focusing on discipline

and the body. Yet he has largely left out the "spatial" in his discussions of religion and power, which I emphasize here as an equally significant element in Foucault (e.g., in *Discipline and Punish* [1977], the panopticon is a disciplinary regime that depends on a particular spatial order as does a military regiment).

12. Harel (2017) notes the recent discursive shift from an active, mystical messianism to a passive, rational embrace of redemption among a younger generation of religious settlers in Alon Shvut. While this is an interesting development, over the course of my fieldwork, I was struck by the fact that the messianic component of ideological settlers (post–Gush Emunim) seemed to have been overstated as a motivating force.

Chapter 1

1. See Attias and Benbassa's (2003:12) discussion of a thematic pattern operating throughout the Bible that features the union of man and his earth, a violation of the law that binds the two, and then a scattering or dispersal from place. They note that God's commandment to Abraham to go forth from his land is one of uprooting and, they aptly call it, "an unfaithfulness to place."

2. Distinct and often competing designations for the Tomb of the Patriarchs—al-Ḥaram al-Ibrahimi (Abraham's Sanctuary) in Arabic and Meʿarat ha-Makhpelah in Hebrew—reveal the degree to which the site is embattled. When referring directly to the settler's relationship to it, I use the Hebrew designation alone. When referring to the site otherwise, I use English and either Hebrew or Arabic or both depending on contexts and the ways I as a researcher encountered them in the field. The other choice is simply enumerating all three designations with each reference, which in my mind shifts attention away from the ways names are used in asymmetrical settings.

3. For a different scholarly reading of this passage, see Michael Fishbane's (2002) discussion regarding *Ḥaye Sarah*. Fishbane talks about the importance of intertextuality in Jewish tradition and the practice, which began in classical Judaism, of reading biblical passages in a fixed cycle in conjunction with other prophetic selections used to suggest parallels and contrasts to the passage in question. The prophetic readings, *haftarot*, he says, use different contents to shape comprehensive analogies by contrast (Fishbane 1998, 2002). For example, the prophetic text that is linked with the passage of Abraham's purchase comes from 1 Kings 1:1–31 (Fishbane 2002:20). It is as follows: King David is old and infirm, and intrigue arises over the issue of succession. Reading the two texts together as is generally done in practice, Fishbane (2002:23) remarks, reveals distinct models of aging. In the Abrahamic type, Abraham is old and prepares for his family succession in an orderly way, while in the Davidic model, there is both physical debility and intrigue as the king battles both at the same time. To expand the implications of the purchase in other possible ways, consider also the halakhic laws that are derived from this passage. They are as follows: when making a land purchase, one must detail its precise features to identify it. This is because in the biblical passage, the Cave of Makhpelah (a double cave) and the toponyms "Kiryat Arba" and "Mamre" are mentioned for clarification. Other areas of Jewish law derived from the purchase include the treatment of those categorized as resident aliens because Abraham himself is an alien, as well as a series of injunctions around burials. These interpretations and religious laws lay the groundwork for an ethics that stands in sharp contrast to the ideological settler position.

4. This discussion is based on a reconstruction of an actual tour with Mageni in the early phases of my fieldwork. In addition to my field notes and recollections, I use another version of the tour "The Road of Valor: From Jerusalem to Hebron with Chaim Mageni" that was transcribed from a recording and published after his death (Mageni Family 2003). Observations about the tour are my own, but are borne out by quotes and textual sources that are recorded in the published version.

5. Todorov (1999:21) reminds us that when Columbus approaches the New World, he never fails to comment on the blackness of skin, on parrots, and on heat; these all seem to be arbitrary observations at first but Columbus sees these as directly linked to wealth—a wealth he is intent on exploiting.

6. Whether one can ever identify a continuous Israelite ethnicity from archaeological evidence is one of the primary debates in the scholarship on archaeology and its relation to nationalism (Kohl 1998; Abu El-Haj 2002). There is also debate about whether kingship in the Bible is mainly fictive or has a historical basis. Yet not even the most biblically oriented of scholars take the patriarchal past of Abraham as the basis of a history that is not mixed with the literary imagination and elements of myth (R. E. Friedman 1997; Hendel 2005).

7. The place name "Mamre" is old and appears, for instance, on the sixth-century Byzantine-era mosaic map that is located in Madaba, Jordan, the oldest surviving cartography of the Holy Land and Jerusalem.

8. Halhoul has a long, multilayered history dating from the Bronze Age. I am commenting on the kinds of claims Mageni makes and what it reveals about the ideological underpinning of religious settlement.

9. The geographer Roger W. Stump (2000:19) makes the point that fundamentalist religious groups often link their form of observance to a particular geographical expression. Control over its boundaries grants a spatial dimension to a moral community, which can be protected from outside threat. Spatial control does not need to happen at the national and territorial level alone; it can also focus on places that have a moral valence. In this case, boundary making with respect to the biblical Land of Israel is an expression of an ideological formation as well as a distinct Jewish settler identity.

10. Between 1993 and 2000, 250 miles of roads were constructed bypassing Palestinian areas (Hanieh 2003).

11. See Breaking the Silence's website, http://www.breakingthesilence.org.il, for its tours and publications.

12. I do not mean to imply that settlers and soldiers always have diametrically opposed points of view. On the contrary, as Chapter 2 (on legalities) shows, religious settlers are increasingly serving in the military to protect settlements on the one hand, and a settler's religious vision is actively being shaped by the ways the military context is being drawn into a religious vision on the other. Shaul's views, though, while not representative of most soldiers, are useful as a way of continuing to explore and contextualize a settler point of view, thinking through some of the elements being masked in an ideological formation.

13. For a more extensive discussion of surveillance and "othering" in the context of the Israeli occupation, see Zureik (2011).

14. This is a very military point of view, that is, understanding the bypass roads and policies of separation as emerging directly out of Palestinian violence and ensuring greater security. Among settler circles, separation was initially more of a far-right idea linked to "transfer" advocated by Rehavam Ze'evi.

15. See http://www.cptpalestine.com for a description of the organization's aims in Hebron.

16. See Braverman (2009) for a discussion of the way political battles in the Israel-Palestine conflict are often fought out through the planting or uprooting of trees.

17. Here I am drawing on ideas in Durkheim's *Elementary Forms of Religious Life* (2008 [1912]) in which there is no inherent content to the sacred apart from its designated opposition to the profane. Boundaries are important, then, in marking off this difference, as is a sacred object's

withdrawal from everyday use. Applying this to the religious settler context, I suggest that, while a settler's attachments to land may be attributed to the Bible alone, a sense of sacredness is actually being reinforced in practice by removing land from its agricultural or daily use in a Palestinian context and designating it instead as a sacred marker of biblical events.

Chapter 2

1. Fahd Kawasmeh later served as mayor of occupied Hebron and was a senior member of the PLO executive committee. In 1980 he was deported to Lebanon by Israel in the wake of an attack on six Hebron settlers for allegedly inciting an atmosphere of violence (Claiborne 1980). In 1985, Yasser Arafat convened the Palestinian National Council in Amman (the Palestinian government in exile), and Kawasmeh was assassinated by a Palestinian faction.

2. For an overview of the relationship between law and morality in ethnography, see Greenhouse (2015).

3. A recently released 1970 document that is signed by Shlomo Gazit confirms the military's role in confiscating private Palestinian land to build the settlement of Kiryat Arba (Berger 2016).

4. Gazit's statement on meeting basic Palestinian needs reveals the paternalism of a so-called liberal occupation stance in which the national aspirations of the Palestinian population did not figure directly into these early administrative endeavors. Gazit also added that Palestinian civilians were seen, from a military point of view, as a logistical problem in comparison with the seemingly greater threat posed by neighboring Arab states, whose armies had previously been at war with Israel. For his written account of the period, see Gazit 1995.

5. Gazit was indirectly referring to the fact that Palestinian Arabs who remained within the 1949 armistice lines of Israel had been granted citizenship but nevertheless remained under military rule until 1966 because they were considered to be a "fifth column" (cf. Robinson 2013).

6. See Berger (2016) for the way the land used to build Kiryat Arba was confiscated by military orders for ostensible security purposes and the housing units on it were initially represented as being for military use.

7. The following discussion is based on the proceedings that took place during the 324th session of the Sixth Knesset (1968), the 49th session of the Seventh Knesset (1970), and the 268th session of the Seventh Knesset (1972).

8. As Walter Benjamin notes (1986:295): "Where frontiers are decided, the adversary is not simply annihilated; indeed he is accorded rights even when the victor's superiority in power is complete." In other words, fixing borders leads to constraints on violence in the aftermath of war.

9. After the signing of Israel's peace agreement with Egypt, the Minister of Agriculture Ariel Sharon used settlement as an opportunity to strengthen Israel's hold on the West Bank.

10. For a classic discussion of the history of this mischaracterization, see Said (1992).

11. The degree to which ideological or religious settlement articulates with or departs from labor-oriented agricultural settlement is part of a heated internal debate, with left and right in stark disagreement. The scholar Gadi Taub (2011), proponent of the ideals of early Zionism, asserts that religious settler claims of continuity with pre-state settlement have essentially rewritten the history of Zionism. What began, in his view, with a focus on state building based on the quest for self-determination ended in religious and ethnic chauvinism. I do not share his view of this untainted period in early Zionism, but nevertheless think that distinct phases of settlement and/or settler ideology need to be charted out and distinguished as a way of historicizing and problematizing the normativity of later claims and practices that take place in areas like Hebron. In other

words, I am not aiming for a full history but a genealogy of an evolving ideological sphere and its relationship to a wider social/legal context.

12. The term *Kach* conveys a sense of militant assertion as a form of explanation in itself. Its visual symbol is a closed fist. In modern Hebrew parlance, the question "why" can be dismissed (or foreclosed) with the response *kakha*, which simply means "it is so." The name *Kach* also carries the sense of a given order and even "truth." I dwell on these various connotations because they are a significant dimension of the party's connection to violence.

Chapter 3

1. Here I am referring to the tension between formerly mobile elements of a post-exilic Judaism (time versus place and marking the body) and a situated origin-based religious settler ethos.

2. Even if one takes the position that reparations were due to the 1929 Jewish community, settler actions foreclosed questions of who would have been its rightful recipients and what forms of reparation were appropriate given the 1948 expulsion of Palestinians from Israel and Palestinian refugees from the 1967 war.

3. For a discussion of the soldier's duty to master emotions as the antithesis of childishness and femininity, see Ben-Ari (2001).

4. *Ma'ariv*, April 17, 1979. Other accounts have slightly different numbers; most frequently, ten mothers and forty children. In any case, children constituted the majority of the residents who lived there for the political aims of their parents.

5. It is interesting to note that, at this early stage, there is no evidence that the Israeli public sees the use of children here as problematic. Since then, however, a discourse has evolved around the recklessness of settlers who place their children in harm's way. In recent protests, women have even been arrested and charged with the endangering their children for including them in protests.

6. For more on the politics of emotion, see Lutz and Abu-Lughod (1990); Stoler (2002); and Ahmed (2004); and for a good overview of affect theory, see Rutherford (2016). On state-sponsored pronatalism in Israel, see Yuval-Davis (1989); Mazali (2002); Sharoni (1997); Kanaaneh (2002); Helman (2011); and Portugese (1998); and on Israeli women and citizenship, see Abdo (2011). On the role of women, maternalism, and dislocation in Palestinian activism, see Jad (1999); Abdo and Lentin (2002); Haj (1992); Hammami (1997); Moors (1995); Peteet (1991); and Slyomovics (1998).

7. Gush Emunim ("Bloc of the Faithful") was a hawkish body and social movement established in the mid-seventies that came out of Israel's National Religious Party (NRP), whose main mission was to promote settlement in the Occupied Territories.

8. Other feminists reading Hannah Arendt, such as Judith Butler (2015), note that the public-private opposition should not be taken as a given but as the result of exclusion and that the terms of exclusion are actually more the ground of the "political" than Arendt admits. Thus, one could read women's maternalist politics here, as some do, as an attempt to challenge assigned gendered roles within the religious community and as emancipatory in its own right. This position, however, disregards the impact of these protests on Palestinians and the forms of exploitation it advances.

9. Lihi Ben Shitrit (2015) sees these protests as "transgressive" and the women involved in them as provisionally stepping out of their assigned gender roles. This raises the important issue of whether to, analytically speaking, consider their protest actions within a traditional ideological frame or beyond it. In this piece, I am concerned with the ways maternal protest uses and temporarily amends traditional gender roles only to reinscribe motherhood in a more entrenched fashion

within the community, though there is no doubt that settler women also participate in less civil and more violent forms of protest not elaborated here.

10. *Eruv* can also mean a ritual Sabbath meal with family and neighbors. Both the physical boundary and meal are interchangeable meanings of *eruv* because they define the social unit or community through activity (Neusner 1991). Outside of settler Judaism, the practice of eating together rather than in a specifically designated place cements the observant community together.

11. For an investigation of the synthesis of liberalism and far-right views among American settlers in the West Bank, see Hirschhorn (2017).

12. Anat Cohen is the radical daughter of the former terrorist Jewish Underground member Moshe Zar.

13. These include "The Pressure Cooker Heats Up," *Haaretz*, April 29, 2001; "Fanning the Flames," *Jerusalem Post*, April 13, 2001; "The Child Sacrifice Debate," *Haaretz*, April 13, 2001, among others.

14. On the state's memorial page to victims of terror, the deaths of other settlers (including David Cohen and Hezzy Mualem) are also recounted.

15. Yitzhak Pas was also front-page news in a feature on Jewish terror (see Fisher 2003).

16. In the early 1990s, Jewish immigration from Israel to the West Bank contributed to most of the population growth in West Bank settlements, whereas by 2011 far more people were born in the settlements than had immigrated to them. In 2011, the total fertility rate (TFR) for Jewish settlers was estimated to be 5.1, whereas that of Israeli Jews was 3.0. The average natural growth (births minus deaths) for 2010, 2011, and 2012 for all Israeli Jews was 1.48 percent, while for the district of Judea and Samaria it was 3.57 percent. See Statistical Abstract of Israel, 2010, 2011, 2012, Table 2.4, cited in Cohen and Gordon (2012). In Kiryat Arba (according to the 2008 Central Bureau of Statistics population census), 42.4 percent of the population was seventeen years or younger and only 6.7 percent was over sixty-five. Missing however, in these statistics is the TFR for those settlers strongly identified with far-right "ideological" views. In 2013, the TFR among Jews in Tel Aviv was 2.58 versus 5.02 for Jews in the West Bank (Judea and Samaria), according to Israel's Central Bureau of Statistics (2013). While these numbers reveal that the total number of births to Jewish settlers in the West Bank was nearly double that of Tel Aviv, it does not provide a clear sense of fertility rates for families living in the most ideological settlements, which tend to be significantly higher.

17. Other than for residents who are already known figures such as Sarah Nachshon and June Leavitt, I have changed the names and details of the lives of people discussed to preserve their anonymity.

18. When messianic ideals are viewed in relation to other features of the ideological landscape, it is evident that the messianic element has been duly overstated, leading to the misperception that it is fervent belief in redemption alone that determines prolific reproduction. As in this conversation, messianism was mentioned as one of several elements that factored into reproductive decisions. See Aviezer Ravitzky (1996) for a comprehensive statement on the issue of messianism; and Janet Aviad (1991) for a similar perspective.

19. The school, Ma'aleh, has made possible a new vehicle of expression for its mainly religious students, many of whom live in West Bank settlements, producing a generation of filmmakers who have sought to take control of the images of religious lives that circulate (Lori 2009).

20. For other anthropological scholarship detailing Israeli bonds to place, see Ben-Ari and Bilu (1997).

Chapter 4

1. For an extended discussion of the way in which spatiality articulates with domination and power, see Foucault (1977, 1980, 1986); and Canetti (1984). While Foucault focuses more on the architecture of power, surveillance built into the order of everyday life, which then gets transposed to the prison (witness the spatial arrangement of the classroom, the factory, and cells of solitary confinement overseen by a panoptical tower), Canetti investigates power in the spatial realm of gesture and movement using prototypical collectivities (crowds or packs) and types of individual rulers. Both the structural and embedded dimensions of spatiality in these discussions of power are important for my representations of settler rallies and celebrations.

2. The alignment between religion and ethnicity in this context resembles Orange parades in Northern Ireland, except that in the Hebron case we have a fundamentalist group linked to an occupation. Sectarianism in Northern Ireland has its own iteration of ethnoreligious settlers but does not make use of the legal conditions of a military occupation where settlers are citizens living among noncitizens beyond the boundaries of the state (cf. Bryan 2000). Finally, Orange parades occur once a year. In Hebron, in contrast, there is an ongoing succession of events that continually brings hundreds and at times thousands of settlers and their supporters to the area. I use the term "ethnicity" here, even though it resembles other racializing processes, because racial justifications tend to be more explicitly biological, physiological, and scientific, whereas in this case, the work of establishing difference and inequality is being accomplished by other means (e.g., militant shows of ethnoreligious belonging) against the backdrop of a legal gray zone. For an opposing view of Israeli ethnicity and parades, see Rosman and Rubel (2015).

3. For an interesting framing of the celebration of Israeli Independence Day within the Green Line, see Virginia Dominguez's (1989: chap. 2) extended discussion "On Ritual and Uncertainty." In this chapter, she discusses the paradox of creating meaningful "rituals" to celebrate this secular holiday when most other collectively observed Jewish holidays already have a prescribed manner of observance. She goes on to describe the display of flags, fireworks, picnics, and play with toy hammers among secular Israelis and the importation of religious prayer and ritual Seder meals among the Modern Orthodox.

4. A strong relationship between the ideological settler community and the military has developed because soldiers serving in elite units are increasingly being recruited from religious settlements. In Hebron, it is common to see soldiers wearing *kippot serugot*, knitted skullcaps worn by settlers and others in the Modern Orthodox camp. As other discussions make clear, soldiers are increasingly being drawn from ideological settlements and often have a strong affinity with the settlers they are protecting. The opposite can also be true, as pullouts of the military make clear. Soldiers just as often resent putting their lives on the line for causes and communities they do not believe in.

5. The idea of private property being upheld as an inviolable principle within a long-standing occupation remains firmly entrenched in the (fluid) legal structure of the occupation to this day. As with other legal limits, there are many ways of circumventing it, and given the other humiliations/violations taking place in this case (soldiers positioned on rooftops, for instance), it seems unusual that it continues to be invoked as a worthy value, even if more in name than in practice. On occasion, the Israeli High Court has even overturned settler or military decisions as violations on these grounds.

6. In her Independence Day discussion, Dominguez (1989: chap. 2) notes the holiday's "awkward secularism." She points out that there has been a tendency to draw on religious forms as a

way of elevating a comparatively secular day to the status of other Jewish holidays. Within Israel the day reflects key Israeli dilemmas over the degree to which the state's military might ought to be put on display and whether to include Palestinian citizens or Palestinian Jerusalemites. In the settlements, however, the religious turn is used to appropriate national symbols for ideological settler aims at the far right end of the spectrum that are overtly exclusive and at times directly opposed to aspects of Israeli nationalism.

7. The state's ethnic character in part derives from the fact that Judaism has been composed of both ethnic and religious elements. In many contexts where Jews are a minority population, the collective has been defined only in part by actual religious observance. A Jewish designation may include language, food, and religious celebrations in addition to having Jewish ancestry. Moreover, intermarriage and conversions have been discouraged, endowing Jews with both a distinct ancestry and a set of cultural markers that are not purely religious. In the United States, for example, Jews who observe Judaism strictly are in the minority (cf. Neusner 2003).

8. Paine's essay too easily creates a continuum between ideological settlers in the contemporary moment and the statements of nineteenth-century Zionists. For this reason, he sees the Goldstein massacre as an exceptional act of violence, distancing the views of other ideological settlers from the violence of Kach and posing the occupation, colonialism, and violence as having little relationship to Jewish settler forms of observance.

9. Ariel Sharon was initially considered a hero. This changed when he decided to oversee the removal of settlers from Gaza in 2005. Though not a religious figure himself, he spearheaded the secular right-wing Likud Party's support of religious settlement and directly translated many of its claims into a secular vernacular. Mirroring the actions of devout settlers in the Old City of Jerusalem, Sharon himself maintained a private residence directly in the Muslim quarter. The massive Israeli flag draped over its portal was supposed to be an assertion of Israeli sovereignty over the historic half of the city that was annexed in 1967 after being captured from the Jordanians.

10. Michael Herzfeld (1993) has explored the social production of indifference within the structure of Western bureaucracy. He points out its relationship to exclusivity and the manner in which it articulates with nationalist logics in distinguishing between insiders and outsiders. In my example here, indifference has the added dimension of orienting settlers through the space of an evolving oppositional geography, producing "difference" as a seemingly timeless primordial quality. Therefore, indifference is a feature of the routinization of dominating spatial practices rather than the trigger that precipitates violence against the other.

11. The concern for environmental "awareness" here is paradoxical at best. Within the settlement the developer is an environmental advocate, protesting among other things against the use of toxic chemicals on lawns. But, in practice, environmental resources such as water can be put to particularistic purposes. For example, excessive water use within Kiryat Arba for lawns and the pool did not figure into his awareness, and the greening of settlement lawns resulted in vast water shortages for Palestinians in Hebron. In the heat of the summers of 1996 and 1999, water reached Palestinian Hebron once every twenty days, while in the settlement there was water to spare.

12. For discussions of the frontier in the Israeli context, see Kimmerling (1983); Kellerman (1993); Yiftachel and Meir (1998).

13. Under Ottoman rule, lands classified as *miri* were state lands leased to tenants with the rights of usufruct, which reverted back if they were left untouched for a period of three years. Thus, an equation, Kimmerling argues, between land ownership and presence emerged that pre-state Zionist settlement deployed as a substitute for actual sovereignty.

14. Years later, this market was shuttered and abandoned because it remained under Israeli

rule after Hebron was split into H1 and H2 areas. Further disconnected from the modern Palestinian city, economically strangled under military rule, and the target of settler activity, the once-vibrant market shut down for many years. It was then reopened within international funding but is still not economically self-sustaining in the face of the settler presence in the Old City.

15. Because of bypass roads, the buses traverse different routes that generally go around Palestinian population centers and are no longer grated. The sense of normalcy gained on the settler side, however, masks years of active confrontation.

16. For a discussion of the relationship between fear and a security regime in Israel, see Ochs (2013).

17. The experience was later reiterated one night coming back on the bus from Jerusalem to Kiryat Arba. Soldiers, in search of a Palestinian suspect during a period of suicide bombings, climbed on board and systematically shined a light into everyone's face without asking any questions or asking for identification and then waved the bus through.

18. This example again predates the H1 and H2 division of authority in Hebron (and the Palestinian Authority in H1), evident from the color of the license plate and the designation on it, though the incident's significance remains relevant to this day.

19. I have also found the following work useful for my formulation of the issue: Ben-Ari and Bilu (1997); Gupta and Ferguson (1997); Lefebvre (1991); Kellerman (1993); and Yiftachel and Meir (1998).

20. Here I am again referencing the language of Robert Paine by using "certainty" (1995).

21. The exception to territorial expansion is the settlement of Yamit, which was dismantled in the wake of the 1982 withdrawal of the Sinai, in accordance with the 1979 Camp David Agreement between Israel and Egypt. Yamit became a touchstone of social protest for the secular right and a rallying cry for the Gush Emunim, which ushered in its own protesters (many of whom still reside in Kiryat Arba) amid considerable media coverage once the original secular inhabitants of the settlement had evacuated the location without protest. Only one other major withdrawal and dismantling of settlements—i.e., Gaza in 2005—has occurred since then.

22. Moledet, whose name literally means "Motherland," was a secular far-right political party, headed by Rehavam Ze'evi (dubbed "Gandhi"), a former general of the Israeli army turned politician, who was later assassinated by the Popular Front for the Liberation of Palestine (PFLP). The party advocated the annexation of the Occupied Territories and forcible "transfer" of all of its Palestinian residents. To provide a more concrete sense of how the "far right" fared in Kiryat Arba, it is worth noting that it captured 21 percent of the vote, second to Likud (27.4) with the National Religious Party (37.4) in the majority. Of the settlement's eligible voters, 82.6 percent participated according to an internal bulletin of the municipal council of Kiryat Arba, *Yedion ha-Matnas*, distributed after the 1996 election. In Kiryat Arba at the time, many of the supporters of Moledet, which cast itself as a nonreligious party, were in fact devout settlers who would have voted for the NRP but thought that the agenda of Palestinian "transfer" outweighed the NRP's more explicitly religious party platform.

23. As we will see in Chapter 5, Baruch Goldstein, a prominent Kiryat Arba resident and member of Kach, was the perpetrator of the 1994 massacre at Meʿarat ha-Makhpelah/al-Ḥaram al-Ibrahimi. Yahya Ayyash, a member of Hamas linked to suicide bombing tactics inside Israel after the massacre, was assassinated in Gaza in 1996 by an exploding cell phone.

24. Settlers of Kiryat Arba were keenly aware that the other half voting for Peres included the so-called Arab vote (Palestinian citizens of Israel), which they asserted would be responsible for determining the political course of the Jewish majority. This sort of phobic rhetoric, which does

not consider Palestinian citizens as true citizens, seems to have been a factor contributing to the boy's error.

25. The negative attitude toward Peres is deeply ironic because he made so many concessions to settlers that were responsible for launching the Gush Emunim.

26. The *hesder* yeshiva, a religious institution of higher learning for men from as young as sixteen to their mid-twenties, is partly responsible for advocating these sorts of interpretations. These yeshivot are distinctive in that they provide both religious education and military training for ideological settlers, as opposed to the ultra-Orthodox camp that has traditionally used religious study in place of fulfilling its military service.

27. This biblical phrase and the title of the essay is telling in itself. In the Bible, Kiryat Arba (literally, the fourth town/village) is being qualified or clarified, so that it reads "Kiryat Arba, which is Hebron," distinguishing it from other villages. The religious settler invocation "Kiryat Arba is Hebron," however, alleges that the Bible states that Kiryat Arba includes all of Hebron. Both are possible readings of the Hebrew, but the context of the biblical passage suggests it is a qualification rather than an inclusive claim.

28. For this reason, ultra-Orthodox groups, including Neterei Karta and Agudat Yisrael, considered much of Zionism's invocation of "return" from exile to be heretical.

29. According to Noahide laws, a *ger toshav* cannot worship idols, murder, steal, or engage in sexual immorality.

30. Gideon Aran, one of the key anthropologists to study the Gush Emunim movement, for instance, characterizes the belief system of its inner circles as structured around a mystic-messianic core that takes over nationalism; see Aran (1990) for a succinct statement of this view.

Chapter 5

1. The account that follows is a synthesis of the Shamgar Commission (1994a, 1994b) report (the government's inquiry in the wake of the massacre), newspaper accounts, interviews, and secondary literature. Meir Shamgar, head of the Shamgar Commission investigating the incident, came from a military background and in 1967 designed the legal infrastructure for the occupying military government. After serving as a judge on the Israeli Supreme Court in the mid-1970s, he served as its president from 1983 to 1995.

2. To provide a sense of these different numbers and accounts of Palestinian fatalities, see Abraham (1994), who discusses bias in mainstream reporting, noting that Jewish settler violence was not a term used in press coverage but rather only as part of an ongoing cycle of violence. His documentation shows that of the twenty-nine killed by Goldstein in the mosque, eight were either children or teenagers. See also Honig-Parnass (1994); Handley (2009), for another overview of media coverage; and Pedahzur and Perliger (2011) and Wiles (2014), for their figures and accounts.

3. The Shamgar Commission (1994b) and news accounts (Struck 1994) report that the testimony of the soldiers Sergeant Kobi Ben-Yosef and Private Nir Drori stationed in the Tomb conflicts with that of Major General Danny Yatom, who claimed that they fired into the air only.

4. For other scholarship on fundamentalism as an essentially psychological orientation, see Strozier et al. (2010).

5. For a more sociological explanation of settler violence as "deviance," see Weisburd (1989); and for a more general, micro-level account that situates violence in social interactions, see Collins (2009).

6. There is also a second kind of religious explanation put forth, which links elements of symbolic violence (and representations of violence) in biblical passages to actual acts of violence

carried out (cf. Horowitz 2008). So for example, the Goldstein massacre took place during Purim, and some scholars argue that the narrative of Purim itself shaped Goldstein's violence in response to impending threat. The wicked minister Haman who plans to kill Queen Esther's uncle Mordechai and annihilate all Persian Jews is instead hung in the gallows he builds. However, using this tale of threat and reversal to understand Goldstein's actions focuses more on "reversal" and settler justifications for the massacre while ignoring the emphasis on "justice restored" in the Bible.

7. In conjunction with this overall exoneration, the floor plan of the mosque in the Shamgar report also indicates some of the ways in which the mosque's space and long history were remapped in relation to this act of violence. For instance, the map points out where Goldstein's body was found, where weaponry was found in relation to the mihrab, or prayer niche, where soldiers were stationed at the separate Jewish entrance and Muslim entrance), and where barriers and a shared door between the Jewish and Muslim side were located.

8. Though the site is deemed sacred to Christians as well, few if any Christians pray there. The contemporary battle over the site is mainly Jewish-Muslim, even though it was a church during Byzantine rule. This adds another piece of evidence to my constructionist argument, namely, that the contemporary sacred stands as a reflection and refraction of the wider social field from which it is fashioned, to again invoke Durkheim on the social basis of the sacred.

9. Writing on the sacred, Hassner (2013) resists what he calls a "poststructuralist" orientation, i.e., the Durkheimian idea that the sacred is but a relational category and that anything can become sacred or be made profane. Following the work of Mircea Eliade (1987), Hassner argues for the sacred's intrinsic characteristics, which he maintains have universal value.

10. There is an extensive literature on sharing the sacred. Anna Bigelow (2010) takes issue with the idea that sharing the sacred generally elicits violence. Studying Indian temples that are sacred to several traditions, she credits the guidance of religious leaders with ensuring harmony, while creating sanctuaries that define themselves against the wider social conflicts taking place outside of them. Leaders can ensure harmony, she mentions, by finding common or overlapping elements within different religious traditions. Yet in the Hebron case, better religious leadership alone would not be sufficient to result in greater harmony. Glenn Bowman (1993) talks about Mar Elias, where Christian and Muslim identities can converge, and the malleability of religious identities in that space. These two views are challenged by Robert Hayden (2013), who, drawing on Bosnia, proposes seeing an underlying power-laden antagonism in shared sacred spaces. The agency of devotees themselves to inscribe sacred spaces with new signs and meanings, as well as the way a site's materiality may prevent or enable broader processes of change, does not substantively enter into these important conversations.

11. Jacob Neusner (personal communication) notes that while Jewish pilgrimage to the graves of the patriarchs or matriarchs cannot be categorically ruled out, the evidence does not support it:

> The locus of pilgrimage in Rabbinic canonical literature, Mishnah through the Babylonian Talmud and Midrash-collections during the first six centuries CE was the following:
>
> 1. Pilgrimage was to Jerusalem/Temple involving animal sacrifice; in the canonical classics there is talk of pilgrimage to Alexandria's Temple and Jerusalem's, but nothing about regular, annual pilgrimages to the graves of the patriarchs or matriarchs. The Christians did that, Jews didn't. Constantine's mother, Helena, was a major player in identifying graves.

2. Graves were visited a year after death for the collection of the bones in ossuaries for secondary burial.

3. There are stories in the Talmud (Bavli) that come to mind about not hanging around graves, which are dangerous places.

12. This shift is not unique but precedes others. In subsequent years, scholars have documented the transformation of Muslim tombs within Israel to sites of Jewish saint worship in a proliferating network of pilgrimage sites (Lehrs 2012; Taub and Sasson 2012).

13. As mentioned earlier, the outer walls of the Tomb of the Patriarchs have been dated to the first century CE by Israeli archaeologists, formally the Herodian period, though it is not certain that the original builder was Herod himself. The existing structure is the product of many subsequent transformations throughout history—Roman, Byzantine, Fatimid, crusader, Ayyubid, Mamluk, and Ottoman. In 1187 after Hebron was recaptured from the crusaders by Salah ad-Din al-Ayyubi (Saladin), the structure again became a mosque, and Saladin is credited with building its minarets, two of which remain standing.

14. Soldiers also testified to the Shamgar Commission that they were not sure of the rules of engagement and whether they could fire on Goldstein to stop him.

15. The concept of "religious pluralism" had been invoked long before the massacre. See, for example, Chaim Herzog's 1976 (A/31/303, S/12223) brief to the UN Security Council as an example of the Israeli government's position emphasizing pluralism rather than Jewish settlement in Hebron, linking it to Jerusalem holy sites: "Israel has administered the West Bank since 1967. The overriding principles guiding its policy regarding all the Holy Places have been, and are, to guarantee the access of members of all faiths to these places and to ensure freedom of worship there to members of every religion. These principles apply to the Tomb of the Patriarchs and thus the sanctity of the shrine is strictly observed."

16. In Jerusalem's Church of the Holy Sepulcher, one also notes an aural competition among various Christian factions, but the distinct history of those divides and competitions says little about remaking the "sacred" and more about the way that space is infused with a distinct set of power dynamics that exist between these related groups. For this reason, I do not see this competition as analogous to the conflict and violence in the Tomb.

17. This gradual transformation of the religious space and the religious sentiments attached to these changes are not without precedent. The 1929 riots throughout the country were set off over European Jewish immigration to Palestine as well as over increased Jewish demands for permanent worship at the Western Wall adjoined to the retaining wall of the Muslim holy site al-Ḥaram al-Sharif.

18. Prior to the 1994 Goldstein massacre, at least two other violent confrontations in the Hall of Isaac came to the attention of authorities. In 1987 and 1988, during the height of the First Intifada, clashes took place there when Muslim worshippers entered the area during the time of the Jewish service, and the military was brought in to intervene.

19. On a visit to the Muslim side of the Tomb with a Palestinian guide, a European tourist asked whether there was any religious reason for the partition. The guide responded: "You see they [Jews] have the memory of Abraham, but Abraham is the father of every religion. You can be a Christian or a Muslim and visit this place. But because Abraham is here, it doesn't mean you can take half the mosque. Take the case of Muslims in Andalusia Spain for instance—they have a lot of history there. It doesn't mean that I go to Alhambra in Andalusia—which also had Jews and Muslims—and occupy Alhambra after eight hundred years." His comments raise the issue of how

to determine the "legitimate" parameters for historical claims as well as whether all biblical claims can be viewed as historical claims.

20. Presumably meaning a lack of Jewish witnesses.

21. Kach, a far-right Orthodox political party that existed from 1971 to 1994, was banned from the Israeli Knesset and considered to be a terrorist organization. Though no longer existing in its original form, its ideas have been resurrected in movements such as the New Kach 2001–2003, and presently in Lehava, which uses violence ostensibly to oppose relationships between Jews and non-Jews or "Arabs" as Meir Kahane did, using the rhetoric of "non-assimilation."

22. Simons (1995) in effect writes an apologetics of the massacre, which he bases on "evidence" from the Israel State Archives and the government's own investigation. His essay is meticulously documented but the logic of the argument confirms this general sense of threat that, if not subscribed to by all settlers, is at least widely regarded as a legitimate security issue.

23. More recently I have seen other settler allegations that Goldstein was set up and that other perpetrators were actually responsible for the massacre, another conspiratorial set of explanations. Yet these are interesting for revealing internal divisions and distrust among different factions of settlers.

24. The historian Jerold S. Auerbach (2009), writing an academically researched defense of Hebron's Jews, takes these explanations of threat at face value in order to construct a defense for Hebron settlement and diminish the problem of settler violence. Yet no academic, Israeli or otherwise, seems to support this view, and it is interesting that he takes settler accounts as a "factual" source of information without seeking other evidence to corroborate it.

25. While the risk for settlers exists, then, their sense of vulnerability requires other explanations. In representing the psychic toll of colonial rule on the colonizer, Memmi (1991:57) concludes: "Deep within, the settler pleads guilty." The more asymmetrical a colonial power structure is, he maintains, the more a certain self-condemnation comes back to haunt those who rule and the more they wish the oppressed would simply disappear (ibid., 97). This internal conflict creates a need to continually account for the distance that exists between the colonizer and the colonized, legitimating colonial privilege through reference to the others' inferiority or lack of character (ibid., 123). The paradox is, as Memmi notes, that the privileged status of a settler depends on there being a population present to colonize. A psychic struggle of this sort, as well as any (real or imagined) shifts in Israeli settlement policy, appears to contribute to a settler sense of vulnerability.

26. The case of Avishai Raviv allegedly provides an example of the resident-informer formula gone awry. Settlers claim that Raviv was allowed to carry on his right-wing activism in exchange for gathering intelligence on residents of Kiryat Arba who were suspected of engaging in criminal activity. They charge that he exploited his position to escape the control of the GSS; specifically, they say Raviv did not inform the GSS about Yigal Amir in spite of the fact that that Amir allegedly told him about the assassination plot beforehand. These early conspiracy theories have not disappeared, as one might expect, but have experienced a revival in right-wing discussions on the Internet.

27. This part of her allegation suggests there is a liberal media and government conspiracy to frame the religious right.

28. The phrase "our public" is frequently used to indicate the religious settler public, as opposed to a general Israeli public. This distinction is itself telling of the ways in which the ideological settler community sees itself positioned within the public sphere of the nation. In fact, it is considered by its settler members to be an entirely distinct group with its own separate interests.

29. *Baruch, ha-Gever* was readily available in Kiryat Arba's public library, even though the book had been officially banned. Its notoriety actually made the book more popular in far-right circles, while preventing public scrutiny and a direct critique of its contents. The title of the book means both "Baruch the man" and "Bless the hero."

30. Paine (1995) paints the settler reorientation of Judaism taking place, as does Dalsheim (2011), as a matter of emphasizing the Land of Israel over other Jewish values. While the concept of land is valued no doubt over that of Torah and people, the shift is far more complicated in that it entails a growing acceptance of violence and the sanctification of perpetrators, as the case of Goldstein demonstrates.

31. Issacharoff and Levinson (2010) documented a memorial ceremony for Goldstein in which Kiryat Arba council members and far-right Kahanist sympathizers together with a large police presence attended and read the book of Esther, invoking parallels between the threat of violence in the Bible and the massacre.

32. Within the last five years, Goldstein sympathizers have allegedly gathered in Kiryat Arba and paraded through the streets of Hebron displaying his picture, provoking Palestinian Hebronites.

33. There is also the recurring question of whether Goldstein acted alone or with accomplices, but, in either case, the larger claim concerning the conditions that led to the massacre still holds true.

34. Reporting in *Haaretz* on the twenty-first anniversary of the Goldstein massacre, Amira Hass (2015) noted that the area around the Tomb remained as deadened as ever, secured by 120 roadblocks and eighteen manned checkpoints. Much of Shuhada Street was off limits to Palestinian pedestrians, while its buildings remained shuttered, making them vulnerable to other acts of settler incursion and takeover.

Chapter 6

1. Barmash (2005) makes the point that the archaeological evidence suggests that the conquered populations of the Northern Kingdom either stayed put in the North or fled south and sought refuge in the Southern Kingdom. The "lost tribes," then, were presumably an early biblical memory invented after the fact to talk about the restoration of a preconquest state rather than an account of what actually took place and it ignored the actual mixing of these two populations.

2. For a recent quest with colonial overtones to recover lost Jews by an adventurer and scholar, see Parfit (2013).

3. "Kuki" is used for this group in India, while "Chin" is used in Burma. More recently the Chin and Kuki have preferred to refer to themselves as "Zo" (cf. Thangtungnung 2014, 2015). "Shinlung" is also used to refer to the group by its mythical origins, a legendary cave in China. There is not yet a written history for the various hill-tribe groups that compose the Kuki-Chin-Mizo apart from oral traditions and missionary accounts.

4. See the website of the Bnei Menashe Council, http://www.bneimenashe.org/page_e.php?id=2, claiming that the population in Israel numbers 2,000, and that another 3,000 are waiting to immigrate there. While it is estimated that there are 3.9 million people who belong to the Kuki-Chin-Mizo group, only a small minority has embraced Judaism, and an even smaller group has immigrated (cf. Haokip 2012).

5. For a detailed ethnographic analysis of this embattled region's various tribal groups, founding myths, and conversion to Christianity, see Weil (2003). Weil states that by the 1980s, Rabbi Eliyahu Avichail had "cemented" the association between the forefather Manasia/Manmasi and

the biblical tribe of Manasseh. However, it remains unclear what the original claim was and how it was reframed to suit standard Jewish narratives by Avichail.

6. For instance, Christian elements were mixed with animistic practices that involved animal sacrifice such as the sacrifice of pigs (and reputedly head-hunting), which was incompatible with Judaism and later excluded from Avichail's retelling of his discovery.

7. The Bnei Menashe loyalty to Rav Avichail was further split with another Israeli successor organization, Shavei Israel, and its leader Michael Freund, who apparently operates quite apart and at odds with the organization Amishav that Avichail established.

8. One might trace this settler preoccupation with origins to the immigrant-settler Zionist project itself, which by definition had a limited relationship to the land in Palestine that Zionists came to settle and colonize. In Zionism and Israeli nationalism this early relationship between people and place was strengthened using scientific discourse and archaeological investigation (Abu El-Haj 2002, 2012). Yet here I point out that the ingathering of "lost tribes" in settlements is not a scientific enterprise but one that depends on legal and religious authority, specifically on the expertise of rabbis as well as the conversion of "lost" members to alleviate doubts about their authenticity. Only later did DNA testing occur, which found no genetic link, at least among men. As Abu El-Haj (2012) points out, the reliance on genetics became significant as a way of rationalizing this enterprise further, but just as interesting to me is the attempt to use science while at the same time denying that science has the final say in what has been deemed by ideological settlers to be a religious matter that subscribes to its own ground of authority.

9. This discussion is a synthesis of various papers (see Neuman 2003, 2004a, 2004b, 2005).

10. In classifying these languages, there is still some question as to whether they have a direct relationship to Chinese. While a variety of dialects have not been extensively studied or well documented, none at all appear to be of Semitic origin.

11. For a discussion of "fragmented" Israeli citizenship in a colonial frontier society, see Shafir and Peled (2002).

12. While the origins of Ethiopian Jewish observance are still a matter of debate, it clearly has a far longer history (even according to theories that it emerged in the fourteenth century out of Christianity) than the more recent claims of the Bnei Menashe.

13. For the controversy elicited by resettling Ethiopians in the West Bank, see Gwertzman (1985).

14. Here she inserts an Ethiopian word (which she claims is "Aarai") and alleges that it means thatched mud hut.

15. Her use of the term "crisis" is a common trope when referring to the difficulty Ethiopians face in assimilating into Israeli culture (cf. Collie 2006).

16. In Ariel, Weiss (2017) notes that former Soviet Union (FSU) settler immigrants are increasingly trapped by their choice to take advantage of the opportunities afforded by living in a settlement. While their vulnerability may at times lead to pragmatism, as she finds, my research suggests that it also serves to reinforce an anti-Arab ideology as a way of fitting into a religious settlement.

17. Remennick (2002) characterizes the Russian immigrant community in Israel as "transnational" insofar as it has remained interested in maintaining cultural and social ties to Russia. This orientation was particularly out of sync with an ideological settler ethos in Hebron, which defined itself as a site of authentic Jewish origin dedicated to negating the exilic sensibility of Jews who had returned. Settlers had expected these new arrivals to cut former ties with the homeland and develop a devout loyalty to their new host community.

18. Remennick (1998) notes that, of the nearly one million people who immigrated from Russia in the 1990s, over a third did not identity as Jewish and over 90 percent were not religiously observant. The immigrants who settled in Kiryat Arba reflected this overall trend.

19. The influx of many non-Jewish Russian immigrants from the former Soviet Union has led to what Krevel-Tovi (2012) refers to as the state's missionary spirit, which focuses on conversion of non-Jews. The state seems to have taken over the concerns of Modern Orthodox rabbis and institutions to proliferate accepted venues of conversion. Krevel-Tovi also makes the connection between conversion and fertility, or the demographic preoccupations of the state, and this certainly applies to the case of settlements as well. Just as prolific reproduction provides one solution to the demographic problem of expanding settlements, so too does immigrant conversion meet the needs of distributing a distinct ethnic population over a site of Jewish origin.

20. While compensation first became a divisive issue after the Oslo Accords, when the Israeli army pulled out of areas of Hebron, it reemerged as an issue in 2000 and 2005 when some veteran settlers and new immigrants felt that their security was at stake. Moreover, during the 2005 withdrawal of settlers from the Gaza Strip, when some refused compensation and others claimed it was too little to cover the cost of starting anew, it was again a divisive issue.

21. Yamit was an agricultural settlement in the Sinai whose residents were well compensated when it was dismantled and the area was returned to Egypt in 1982. It then became a dramatic site of protest orchestrated by the Gush Emunim, who stepped in after its former residents had already left. These protests have become etched in the national imaginary as dramatizing the existential dilemmas of settler withdrawal.

22. In 2012, the Jerusalem District Court held the settler purchase of the building valid, and six years after its initial decision, the High Court reversed itself and upheld the lower court's ruling allowing settlers to move back in (Lazaroff 2014). This is but another an example of illegal actions being legalized retroactively by the High Court for settler political aims, and it runs counter to Yuri's narrative.

Conclusion

1. Ten years after the pullout, most Israelis view it as a mistake, strengthening the ideological settler resolve to not make compromises on the West Bank (see Booth and Eglash 2015).

2. The average compensation of Gazan settlers amounted to approximately $200,000–$300,000 per family if accepted before the pullout (*New York Times* 2005).

3. For a further discussion of belonging and exclusion focused on Bedouin residents in the Negev, see McKee (2015).

4. For an overview of the dilemmas of conducting ethnographic research in unstable places, see Greenhouse (2002).

References

Abbady, Tal. 2015. "Ethiopian Jews Recall Israel." *Sun-Sentinel*, December 26. http://articles.sun-sentinel.com/2005-12-26/news/0512250125_1_ethiopian-jews-ethiopian-children-israel.

Abdo, Nahla. 2011. *Women in Israel: Race, Gender and Citizenship*. London: Zed Books.

Abdo, Nahla, and Ronit Lentin. 2002. *Women and the Politics of Military Confrontation: Palestinian and Israeli Gendered Narratives of Dislocation*. New York: Berghahn Books.

Abraham, Nabeel. 1994. "What About the Victims?" *Lies of Our Times*. May. http://cosmos.ucc.ie/cs1064/jabowen/IPSC/articles/article0002775.html.

Abu El-Haj, Nadia. 2002. *Facts on the Ground: Archaeological Practice and Territorial Self-Fashioning in Israeli Society*. Chicago: University of Chicago Press.

———. 2012. *The Genealogical Science: The Search for Jewish Origins and the Politics of Epistemology*. Chicago: University of Chicago Press.

Ahmed, Sara. 2004. *The Cultural Politics of Emotion*. New York: Routledge.

AlSayyad, Nezar, and Mejgan Massoumi, eds. 2011. *The Fundamentalist City? Religiosity and the Remaking of Urban Space*. New York: Routledge.

Andoni, Lamis. 1997. "Redefining Oslo: Negotiating the Hebron Protocol." *Journal of Palestine Studies* 26 (3): 17–30.

Appadurai, Arjun. 1996. *Modernity at Large: Cultural Dimensions of Globalization*. Minneapolis: University of Minnesota Press.

Applegate, Celia. 1990. *A Nation of Provincials: The German Idea of Heimat*. Berkeley: University of California Press.

Aran, Gideon. 1990. "Redemption as Catastrophe: The Gospel of Paradox." In *Religious Radicalism and Politics in the Middle East*, edited by Emmanuel Sivan and Menachem Friedman, 157–76. Albany: State University of New York Press.

———. 1991. "Jewish Zionist Fundamentalism: The Bloc of the Faithful in Israel (Gush Emunim)." In *Fundamentalisms Observed*, edited by Martin E. Marty and R. Scott Appleby, 265–344. Chicago: University of Chicago Press.

———. 2013. *Kookism: The Roots of Gush Emunim, the Settlers' Culture, Zionist Theology and Messianism in Our Age*. Jerusalem: Carmel Press. [In Hebrew]

Arendt, Hannah. 1970. *On Violence*. New York: Mariner Books.

———. 1998. *The Human Condition*. Chicago: University of Chicago Press.

Asad, Talal. 1993. *Genealogies of Religion: Discipline and Reasons of Power in Christianity and Islam*. Baltimore: Johns Hopkins University Press.

———. 2003. *Formations of the Secular: Christianity, Islam, Modernity*. Stanford, CA: Stanford University Press.

References

Attias, Jean-Christophe, and Esther Benbassa. 2003. *Israel, the Impossible Land*. Stanford, CA: Stanford University Press.

Auerbach, Jerold S. 2009. *Hebron Jews: Memory and Conflict in the Land of Israel*. Lanham, MD: Rowman & Littlefield.

Aviad, Janet O'Dea. 1983. *Return to Judaism: Religious Renewal in Israel*. Chicago: University of Chicago Press.

———. 1991. "The Messianism of Gush Emunim." In *Jews and Messianism in the Modern Era: Metaphor and Meaning*, edited by Jonathan Frankel, 197–216. Studies in Contemporary Jewry 7. Oxford: Published for the Institute of Contemporary Jewry, Hebrew University of Jerusalem by Oxford University Press.

Avichail, Eliyahu. 1990. *The Tribes of Israel: The Lost and the Dispersed*. Jerusalem: Amishav.

———. 1998. "How I Happened to 'Discover' the Bnei Menashe." In *Jews in Places You Never Thought Of*, edited by Karen Primack, 31–37. Hoboken, NJ: KTAV Publishing House.

Avishai, Orit. 2008. "'Doing Religion' in a Secular World: Women in Conservative Religions and the Question of Agency." *Gender & Society* 22 (4): 409–33.

Bar'el, Zvi. 2008. "Ghetto Mentality." *Haaretz*, April 17. http://www.haaretz.com/ghetto-mentality-1.244141.

Barmash, Pamela. 2005. "At the Nexus of History and Memory: The Ten Lost Tribes." *AJS Review* 29 (2): 207–36.

Barth, Fredrik. 1998. *Ethnic Groups and Boundaries: The Social Organization of Culture Difference*. Long Grove, IL: Waveland Press.

Beinin, Joel, and Rebecca L. Stein. 2006. *The Struggle for Sovereignty: Palestine and Israel, 1993–2005*. Stanford, CA: Stanford University Press.

Ben-Ari, Eyal. 2001. *Mastering Soldiers: Conflict, Emotions, and the Enemy in an Israeli Military Unit*. New York: Berghahn Books.

Ben-Ari, Eyal, and Yoram Bilu, eds. 1997. *Grasping Land: Space and Place in Contemporary Israeli Discourse and Experience*. Albany: State University of New York Press.

Ben-Dor Benite, Zvi. 2009. *The Ten Lost Tribes: A World History*. New York: Oxford University Press.

Ben-Horin, Michael., ed. 1995. *Baruch Ha-Gever [Baruch, the Man]: A Memorial Volume for Dr. Baruch Goldstein*. Jerusalem: special publication.

Benjamin, Walter. 1986. *Reflections: Essays, Aphorisms, Autobiographical Writings*. Edited by Peter Demetz. New York: Schocken.

Ben Shitrit, Lihi. 2013. "Women, Freedom, and Agency in Religious Political Movements." *Journal of Middle East Women's Studies* (Indiana University Press) 9 (3): 81–107.

———. 2015. *Righteous Transgressions: Women's Activism on the Israeli and Palestinian Religious Right*. Princeton, NJ: Princeton University Press.

Berger, Yotam. 2016. "Secret 1970 Document Confirms First West Bank Settlements Built on a Lie." *Haaretz*, July 28. http://www.haaretz.com/israel-news/.premium-1.733746.

Bernis, Jonathan. 2009. *The Lost Tribes of Israel: Beta Avraham & Beta Israel of Ethiopia and Bnei Menashe of India*. Phoenix, AZ: Jewish Voice Ministries International.

Bigelow, Anna. 2010. *Sharing the Sacred: Practicing Pluralism in Muslim North India*. Oxford: Oxford University Press.

Booth, William, and Ruth Eglash. 2015. "A Decade Later, Many Israelis See Gaza Pullout as a Big Mistake." *Washington Post*, August 15.

Boudreau, Geneviève Boucher. 2014. "Radicalization of the Settlers' Youth: Hebron as a Hub for Jewish Extremism." *Global Media Journal: Canadian Edition* 7 (1): 69–85.

References

Bourdieu, Pierre. 1977. *Outline of a Theory of Practice.* Cambridge: Cambridge University Press.

Bowman, Glenn. 1993. "Nationalizing the Sacred: Shrines and Shifting Identities in the Israeli-Occupied Territories." *Man*, n.s., 28 (3): 431–60.

Brauch, Julia, Anna Lipphardt and Alexandra Nocke, eds. 2008. *Jewish Topographies.* Aldershot: Ashgate.

Braverman, Irus. 2009. *Planted Flags: Trees, Land, and Law in Israel/Palestine.* New York: Cambridge University Press.

Brinkley, Joel. 1991. "200 Ethiopians Trapped in West Bank." *New York Times*, June 6, sec. World. http://www.nytimes.com/1991/06/06/world/200-ethiopians-trapped-in-west-bank.html.

Brown, Wendy. 2013. "Introduction." In *Is Critique Secular?: Blasphemy, Injury, and Free Speech*, edited by Talal Asad, Wendy Brown, Judith Butler, and Saba Mahmood, 1–13. New York: Fordham University Press.

Bryan, Dominic. 2000. *Orange Parades: The Politics of Ritual, Tradition and Control.* London: Pluto Press.

Butler, Judith. 2015. *Notes Toward a Performative Theory of Assembly.* Cambridge, MA: Harvard University Press.

Byman, Daniel, and Natan Sachs. 2012. "The Rise of Settler Terrorism." *Foreign Affairs*, August 18. https://www.foreignaffairs.com/articles/israel/2012-08-18/rise-settler-terrorism.

Campos, Michelle. 2007. "Remembering Jewish-Arab Contact and Conflict." In *Reapproaching Borders: New Perspectives on the Study of Israel Palestine*, edited by Sandy Sufian and Mark Levine, 41–66. New York: Rowman & Littlefield.

Canetti, Elias. 1984. *Crowds and Power.* Trans. Carol Stewart. New York: Farrar, Straus and Giroux.

Casey, Edward S. 1993. *Getting Back into Place: Toward a Renewed Understanding of the Place-World.* Bloomington: Indiana University Press.

Central Bureau of Statistics [Israel]. 2013. "Fertility Rates by District, Sub-District, and Mothers Population Group, Religion, and Age." http://www.cbs.gov.il/publications16/1632_live_birth_2013/pdf/tab07.pdf.

———. 2016. "Population and Density per Sq. Km. in Localities Numbering 5,000 Residents and More on 31.12.2015(1)." http://www.cbs.gov.il/reader/shnaton/templ_shnaton_e.html?num_tab=st02_24&CYear=2016.

Claiborne, William. 1980. "Israeli Court Challenges Deportations." *Washington Post*, May 21. https://www.washingtonpost.com/archive/politics/1980/05/21/israeli-court-challenges-deportations/9a84eb46-bcca-40b3-a8e8-967c8c7115af/.

Clifford, James. 1994. "Diasporas." *Cultural Anthropology* 9 (3): 302–38.

Cohen, Hillel. 2015. *Year Zero of the Arab-Israeli Conflict, 1929.* Trans. Haim Watzman. Waltham, MA: Brandeis University Press.

Cohen, Yinon, and Neve Gordon. 2012. "The Demographic Success of the West Bank Settlements." https://tinyurl.com/hc7e9yr.

Collie, Tim. 2006. "Help on the Way for Ethiopian Jews in Israel." *Sun-Sentinel*, January 29. http://articles.sun-sentinel.com/2006-01-29/news/0601280158_1_ethiopian-jews-ethiopian-national-project-israel.

Collins, Randall. 2009. *Violence: A Micro-Sociological Theory.* Princeton, NJ: Princeton University Press.

Cooper, Alanna E. 2006. "Conceptualizing Diaspora: Tales of Jewish Travelers in Search of the Lost Tribes." *AJS Review* 30 (1): 95–117.

Cresswell, Tim. 1996. *In Place/Out of Place: Geography, Ideology, and Transgression*. Minneapolis: University of Minnesota Press.

———. 2004. *Place: A Short Introduction*. Malden, MA: Wiley-Blackwell.

Curtius, Mary. 1994. "Lost Tribe Applicants Stir Debate on Israel Citizenship." *Los Angeles Times*, August 29. http://articles.latimes.com/1994-08-29/news/mn-32563_1_lost-tribe.

Dalsheim, Joyce. 2011. *Unsettling Gaza: Secular Liberalism, Radical Religion, and the Israeli Settlement Project*. New York: Oxford University Press.

de Certeau, Michel. 1984. *The Practice of Everyday Life*. Trans. Steven Rendall. Berkeley: University of California Press

Demant, Peter Robert. 1988. "Ploughshares into Swords: Israeli Settlement Policy in the Occupied Territories, 1967–1977." Ph.D. dissertation, University of Amsterdam.

Dominguez, Virginia R. 1989. *People as Subject, People as Object: Selfhood and Peoplehood in Contemporary Israel*. Madison: University of Wisconsin Press.

Durkheim, Emile. 2008[1912]. *The Elementary Forms of Religious Life*. Trans. Carol Cosman. Oxford: Oxford University Press.

Egorova, Julia. 2015. "Redefining the Converted Jewish Self: Race, Religion, and Israel's Bene Menashe." *American Anthropologist* 117 (3): 493–505.

Eichner, Itamar. 2010. "Members of Bnei Menashe to Make Aliyah." *Ynetnews*, August 1. http://www.ynetnews.com/articles/0,7340,L-3831308,00.html.

Eliade, Mircea. 1987. *The Sacred and the Profane: The Nature of Religion*. Trans. Willard R. Trask. New York: Harcourt.

Elkins, Caroline, and Susan Pedersen, eds. 2005. *Settler Colonialism in the Twentieth Century: Projects, Practices, Legacies*. New York: Routledge.

Elon, Amos. 1997. "Far City in the Fog." Chap. 5 in *A Blood-Dimmed Tide: Dispatches from the Middle East*. Appeared in *Haaretz*, February 10, 1979. New York: Columbia University Press.

El-Or, Tamar. 2002. *Next Year I Will Know More: Literacy and Identity Among Young Orthodox Women in Israel*. Detroit: Wayne State University Press.

El-Or, Tamar, and Gideon Aran. 1995. "Giving Birth to a Settlement: Maternal Thinking and Political Action of Jewish Women on the West Bank." *Gender & Society* 9 (1): 60–78.

Ettinger, Yair. 2012. "Israel's High Court Affirms That Thousands of Jewish Conversions Are Kosher." *Haaretz*, April 27.

"Ex-Undercover Agent Charged as a Link in Rabin Killing." 1999. *New York Times*, April 26. http://www.nytimes.com/1999/04/26/world/ex-undercover-agent-charged-as-a-link-in-rabin-killing.html.

Falk, Richard. 2013. "Oslo +20: A Legal Historical Perspective." In *The Oslo Accords, 1993–2013: A Critical Assessment*, edited by Petter Bauck and Mohammed Omer, 63–74. Cairo: American University in Cairo Press.

Feige, Michael. 2001. "Jewish Settlement of Hebron: The Place and the Other." *GeoJournal* 53 (3): 323–33.

———. 2009. *Settling in the Hearts: Jewish Fundamentalism in the Occupied Territories*. Detroit: Wayne State University Press.

———. 2016. "Jewish Ideological Killers: Religious Fundamentalism or Ethnic Marginality?" In *Contemporary Israel: New Insights and Scholarship*, edited by Frederick E. Greenspahn, 166–92. New York University Press.

Feldman, Allen. 1991. *Formations of Violence: The Narrative of the Body and Political Terror in Northern Ireland*. Chicago: University of Chicago Press.

References

Fenster, Tovi. 1998. "Ethnicity, Citizenship, Planning and Gender: The Case of Ethiopian Immigrant Women in Israel." *Gender, Place and Culture: A Journal of Feminist Geography* 5 (2): 177–89.

Fishbane, Michael. 1998. *The Exegetical Imagination: On Jewish Thought and Theology.* Cambridge, MA: Harvard University Press.

———. 2002. *The JPS Bible Commentary: Haftarot.* Bilingual ed. Philadelphia: Jewish Publication Society.

———. 2012. *The Midrashic Imagination: Jewish Exegesis, Thought, and History.* Albany, NY: SUNY Press.

Fisher, Ian. 2003. "Israelis Worry About Terror, by Jews Against Arabs." *New York Times,* August 19.

Fonrobert, Charlotte Elisheva. 2009. "The New Spatial Turn in Jewish Studies." *AJS Review* 33 (1): 155–64.

Foucault, Michel. 1977. *Discipline and Punish: The Birth of the Prison.* New York: Vintage.

———. 1980. "Questions on Geography." In *Power/Knowledge: Selected Interviews and Other Writings, 1972–1977,* edited by Colin Gordon. New York: Pantheon Books.

———. 1986. "Of Other Spaces." *Diacritics* 16 (1): 22–27.

Friedland, Roger, and Richard Hecht. 2000. *To Rule Jerusalem.* Berkeley: University of California Press.

Friedman, Richard Elliott. 1997. *Who Wrote the Bible?* New York: Harper One.

Friedman, Shimi. 2015. "Hilltop Youth: Political-Anthropological Research in the Hills of Judea and Samaria." *Israel Affairs* 21 (3): 391–407.

Friedman, Thomas L. 1985. "Ethiopians Are a Joy and a Challenge for Israelis." *New York Times,* January 13, sec. Week in Review. http://www.nytimes.com/1985/01/13/weekinreview/ethiopians-are-a-joy-and-a-challenge-for-israelis.html.

Friedrich, Paul. 1989. "Language, Ideology, and Political Economy." *American Anthropologist* 91 (2): 295–312.

Gazit, Shlomo. 1995. *The Carrot and the Stick: Israel's Policy in Judaea and Samaria, 1967–68.* Washington, DC: B'nai B'rith Books.

———. 2003. *Trapped Fools: Thirty Years of Israeli Policy in the Territories.* New York: Routledge.

Geertz, Clifford. 1973. *The Interpretation of Cultures.* New York: Basic Books.

Glickman, Aviad. 2008. "High Court Orders Disputed House in Hebron Vacated." *Ynetnews,* November 16. http://www.ynetnews.com/articles/0,7340,L-3623612,00.html.

Goffman, Erving. 1986. *Frame Analysis: An Essay on the Organization of Experience.* Foreword by Bennett Berger. Boston: Northeastern University Press.

Golan, Galia. 2013. "Peace Plans, 1993–2012." In *Routledge Handbook on the Israeli-Palestinian Conflict,* edited by Joel Peters and David Newman, 92–106. New York: Routledge.

Gonen, Amiram. 2004. "Homecoming at Burial." *Horizons in Geography* nos. 60–61 (January): 421–26.

Greenhouse, Carol J. 2002. "Introduction: Altered States, Altered Lives." In *Ethnography in Unstable Places: Everyday Lives in Contexts of Dramatic Political Change,* edited by Elizabeth Mertz, Kay B. Warren, and Carol Greenhouse, 1–36. Durham, NC: Duke University Press.

———. 2015. "Law." In *A Companion to Moral Anthropology,* edited by Didier Fassin, 432–48. Hoboken, NJ: John Wiley & Sons.

Gupta, Akhil, and James Ferguson. 1997. *Culture, Power, Place: Explorations in Critical Anthropology.* Durham, NC: Duke University Press.

Gutman, Matthew. 2003. "Father of Shalhevet Pass Arrested for Crimes Against Palestinians." *Jerusalem Post*, July 20.
Gwertzman, Bernard. 1985. "Ethiopian Jews Said to Resettle on the West Bank." *New York Times*, January 18, sec. World. http://www.nytimes.com/1985/01/18/world/ethiopian-jews-said-to-resettle-on-west-bank.html.
Haj, Samira. 1992. "Palestinian Women and Patriarchal Relations." *Signs* 17 (4): 761–78.
Halkin, Hillel. 2002. *Across the Sabbath River: In Search of a Lost Tribe of Israel*. Boston: Houghton Mifflin Harcourt.
Handel, Ariel, Marco Allegra, and Erez Maggor, eds. 2017. *Normalizing Occupation: The Politics of Everyday Life in the West Bank Settlements*. Bloomington: Indiana University Press.
Hammami, Rema. 1997. "Palestinian Motherhood and Political Activism on the West Bank and Gaza Strip." In *The Politics of Motherhood: Activist Voices from Left to Right*, edited by Alexis Jetter, Annelise Orleck, and Diana Taylor, 161–68. Hanover, NH: University Press of New England.
Hammami, Rema, and Salim Tamari. 2001. "The Second Uprising: End or New Beginning?" *Journal of Palestine Studies* 30 (2): 5–25.
Hanafi, Sari. 2009. "Spacio-cide: Colonial Politics, Invisibility and Rezoning in Palestinian Territory." *Contemporary Arab Affairs* 2 (1): 106–21.
Handley, Robert L. 2009. "The Conflicting Israeli-Terrorist Image." *Journalism Practice* 3 (3): 251–67.
Hanieh, Adam. 2003. "A Road Map to the Oslo Cul-de-Sac." Middle East Research and Information Project, May 15. http://www.merip.org/mero/mero051503.
Haokip, George T. 2012. "Jewish Identity in North East India." *Eastern Anthropologist* 65 (2): 181–92.
Harel, Assaf. 2017. "Beyond Gush Emunim: On Contemporary Forms of Messianism Among Religiously Motivated Settlers in the West Bank." In *Normalizing Occupation: The Politics of Everyday Life in the West Bank Settlements*, edited by Ariel Handel, Marco Allegra, and Erez Maggor, 128–50. Bloomington: Indiana University Press.
Harvey, David. 1993. "From Space to Place and Back Again: Reflections on the Condition of Postmodernity." In *Mapping the Futures: Local Cultures, Global Change*, edited by Jon Bird, Barry Curtis, Tim Putnam, and Lisa Tickner. London: New York: Routledge.
Hass, Amira. 2002. "Israel's Closure Policy: An Ineffective Strategy of Containment and Repression." *Journal of Palestine Studies* 31 (3): 5–20.
———. 2015. "21 Years After Goldstein Massacre, Once-Thriving Hebron Is a Mere Memory." *Haaretz*, February 21. http://www.haaretz.com/israel-news/.premium-1.643500.
Hassner, Ron E. 2013. *War on Sacred Grounds*. Ithaca, NY: Cornell University Press.
Hayden, Robert M. 2013. "Intersecting Religioscapes and Antagonistic Tolerance: Trajectories of Competition and Sharing of Religious Spaces in the Balkans." *Space & Polity* 17 (3): 320–34.
Helman, Sara. 1999. "From Soldiering and Motherhood to Citizenship: A Study of Four Israeli Peace Protest Movements." *Social Politics: International Studies in Gender, State & Society* 6 (3): 292–313.
———. 2011. "'Let Us Help Them to Raise Their Children into Good Citizens': The Lone-Parent Families Act and the Wages of Care-Giving in Israel." *Social Politics: International Studies in Gender, State & Society* 18 (1): 52–81. doi: 10.1093/sp/jxr004.
Hendel, Ronald. 2005. *Remembering Abraham*. Oxford: Oxford University Press.
Herzfeld, Michael. 1993. *The Social Production of Indifference: Exploring the Symbolic Roots of Western Bureaucracy*. Chicago: University of Chicago Press.

References

Herzog, Esther. 2007. "Image and Reality in an Israeli 'Absorption Center' for Ethiopian Immigrants." In *Professional Identities: Policy and Practice in Business and Bureaucracy*, edited by Shirley Ardener and Fiona Moore. New York: Berghahn Books.

Herzog, Hanna. 2009. "Imagined Communities: State, Religion, and Gender in Jewish Settlements." In *Gendering Religion and Politics: Untangling Modernities*, edited by Hanna Herzog and Ann Braude. New York: Palgrave Macmillan.

Hirschhorn, Sara Yael. 2017. *City on a Hilltop: American Jews and the Israeli Settler Movement*. Cambridge, MA: Harvard University Press.

Ho, Engseng. 2006. *The Graves of Tarim: Genealogy and Mobility Across the Indian Ocean*. Berkeley: University of California Press.

Hobsbawm, Eric, and Terence Ranger, eds. 1983. *The Invention of Tradition*. Cambridge: Cambridge University Press.

Honig-Parnass, Tikva. 1994. "Palestinian and Israeli Reactions to the Massacre." *News from Within* 10 (4): 2–5.

Horowitz, Elliott S. 2008. *Reckless Rites: Purim and the Legacy of Jewish Violence*. Princeton, NJ: Princeton University Press.

Huyssen, Andreas. 2003. "Diaspora and Nation: Migration into Other Pasts." *New German Critique*, no. 88 (January): 147–64.

Inbari, Motti. 2009. *Jewish Fundamentalism and the Temple Mount: Who Will Build the Third Temple?* Albany: State University of New York Press.

———. 2012. *Messianic Religious Zionism Confronts Israeli Territorial Compromises*. Cambridge: Cambridge University Press.

Ir Amim and Keshev. 2013. "Dangerous Liaison: The Dynamics of the Rise of the Temple Movements and Their Implications." Jerusalem: Ir Amim and Keshev. http://www.ir-amim.org.il/sites/default/files/Dangerous%20Liaison_0.pdf.

Ish-Shalom, Benjamin. 1993. *Rav Avraham Itzhak HaCohen Kook: Between Rationalism and Mysticism*. Albany: State University of New York Press.

Israeli Committee Against House Demolitions. 2013. "Dozens of Settlers Rip Up Atta Jaber's Plants—Under Nose of Police and Kiryat Arba Swimming Pool." Mondoweiss.net, July. http://mondoweiss.net/2013/07/dozens-of-settlers-rip-up-atta-jabers-plants-under-nose-of-police-and-kiryat-arba-swimming-pool.

Israel Ministry of Foreign Affairs. 2015. "The Israeli-Palestinian Interim Agreement." Accessed August 19. http://www.mfa.gov.il/mfa/foreignpolicy/peace/guide/pages/the%20israeli-palestinian%20interim%20agreement.aspx.

Issacharoff, Avi, and Chaim Levinson. 2010. "Settlers Remember Gunman Goldstein; Hebron Riots Continue." *Haaretz*, February 28. http://www.haaretz.com/settlers-remember-gunman-goldstein-hebron-riots-continue-1.263834.

Jad, Islah. 1999. "From Salons to the Popular Committees: Palestinian Women, 1919–1989." In *The Israel/Palestine Question*, edited by Ilan Pappe, 125–42. New York: Routledge.

Jetter, Alexis, Annelise Orleck, and Diana Taylor, eds. 1997. *The Politics of Motherhood: Activist Voices from Left to Right*. Hanover, NH: University Press of New England.

Kanaaneh, Rhoda. 2002. *Birthing the Nation*. Berkeley: University of California Press.

Karp, Yehudit. 1984. *The Karp Report: An Israeli Government Inquiry into Settler Violence Against Palestinians on the West Bank*. Washington, DC: Institute for Palestine Studies.

Kellerman, Aharon. 1993. *Society and Settlement: Jewish Land of Israel in the Twentieth Century*. Albany: State University of New York Press.

Kelly, Tobias. 2006. "Documented Lives: Fear and the Uncertainties of Law During the Second Palestinian Intifada." *Journal of the Royal Anthropological Institute* 12 (1): 89–107.

———. 2009. *Law, Violence and Sovereignty Among West Bank Palestinians*. Cambridge: Cambridge University Press.

Kertzer, David. 1979. "Gramsci's Concept of Hegemony: The Italian Church-Communist Struggle." *Dialectical Anthropology* 4 (4): 321–28.

Khalidi, Rashid. 2006. *The Iron Cage: The Story of the Palestinian Struggle for Statehood*. Boston: Beacon Press.

———. 2013. *Brokers of Deceit: How the U.S. Has Undermined Peace in the Middle East*. Boston: Beacon Press.

Kimmerling, Baruch. 1983. *Zionism and Territory: The Socio-Territorial Dimensions of Zionist Politics*. Berkeley: Institute of International Studies, University of California.

Kirsch, Stuart. 1997. "Lost Tribes: Indigenous People and the Social Imaginary." *Anthropological Quarterly* 70 (2): 58–67.

Klein, Menachem. 2014. *Lives in Common: Arabs and Jews in Jerusalem, Jaffa and Hebron*. Oxford: Oxford University Press.

Knesset Proceedings. 1968. "The Deportation of Hebron Settlers and Discussion." 6th Knesset, 324th Session, August 14: 3318–3327. Jerusalem. [In Hebrew]

———. 1970. "The Founding of Kiryat Arba." 7th Knesset, 49th Session, March 9: 1070–1074. Jerusalem. [In Hebrew]

———. 1972. "The Continuation of the Development of Kiryat Arba-Hebron." 7th Knesset, 268th Session: 1321–1332. Jerusalem. [In Hebrew]

Knott, Kim. 2005. *The Location of Religion: A Spatial Analysis*. London: Equinox.

———. 2009. "From Locality to Location and Back Again: A Spatial Journey in the Study of Religion." *Religion* 39 (2): 154–60.

Koffman, David S. 2011. "The Jews' Indian: Native Americans in the Jewish Imagination and Experience, 1850–1950." Ph.D. dissertation, New York University.

Kohl, Philip L. 1998. "Nationalism and Archaeology: On the Constructions of Nations and the Reconstructions of the Remote Past." *Annual Review of Anthropology* 27 (1): 223–46. doi:10.1146/annurev.anthro.27.1.223.

Kook, Zevi Judah ben Abraham Isaac. 1991. *Torat Eretz Yisrael: The Teachings of HaRav Tzvi Yehuda HaCohen Kook*. Sihot Ha-Rav Tsevi Yehudah. English. Selections. Jerusalem: Torat Eretz Yisrael Publications.

Kravel-Tovi, Michal. 2012. "'National Mission': Biopolitics, Non-Jewish Immigration and Jewish Conversion Policy in Contemporary Israel." *Ethnic and Racial Studies* 35 (4): 737–56.

Krieger, Hilary Leila. 2006. "Bnei Menashe Aliya, Conversions Halted Pending Government Review." *Jerusalem Post*, July 2. http://www.jpost.com/Israel/Bnei-Menashe-aliya-conversions-halted-pending-government-review.

Lazaroff, Tovah. 2014. "HJC Authenticates Jewish Purchase of Beit HaShalom in Hebron." *Jerusalem Post*, March 11. http://www.jpost.com/National-News/HJC-authenticates-Jewish-purchase-of-Beit-HaShalom-in-Hebron-345049.

Leavitt, June. 2002. *Storm of Terror: A Hebron Mother's Diary*. Chicago: Ivan R. Dee.

Lebel, Udi. 2016. "The 'Immunized Integration' of Religious-Zionists Within Israeli Society: The Pre-Military Academy as an Institutional Model." *Social Identities: Journal for the Study of Race, Nation and Culture* 22 (6): 642–60.

Lefebvre, Henri. 1991. *The Production of Space*. Oxford, UK: Blackwell.

Lehrs, Lior. 2012. "Political Holiness: Negotiating Holy Places in Eretz Israel/Palestine, 1937–2003." In *Sacred Space in Israel and Palestine: Religion and Politics*, edited by Marshall J. Breger, Yitzhak Reiter, and Leonard Hammer, 228–50. New York: Routledge.

Leon, Nissim. 2015. "Self-Segregation of the Vanguard: Judea and Samaria in the Religious-Zionist Society." *Israel Affairs* 21 (3): 348–60.

Levy, André. 2001. "Center and Diaspora: Jews in Late-Twentieth-Century Morocco." *City & Society* 13 (2): 245–70.

Levy, André, and Alex Weingrod, eds. 2004. *Homelands and Diasporas: Holy Lands and Other Places*. Stanford, CA: Stanford University Press.

Lior, Ilan. 2015. "Minister Shalom Pledges to Bring More Bnei Menashe and Falash Mura to Israel." *Haaretz*, June 25. http://www.haaretz.com/israel-news/.premium-1.663057.

Litani, Yehuda. 1979. "Tsefardeʻa u-Shemah Hadasah [A Frog and Its Name Is Hadassah]." *Haaretz*, June 15.

Livneh, Neri. 2002. "Coming Home." *Haaretz*, July 18.

Lomnitz, Claudio. 2001. *Deep Mexico, Silent Mexico: An Anthropology of Nationalism*. Minneapolis: University of Minnesota Press.

Lori, Aviva. 2009. "Aḥare she-Nilḥemah ʻal Betah be-Hitnaḥalut Śa-Nur Biyemah Miryam Adler Sereṭ Merots ha-Yeludah ba-Ḥevrah ha-Datit" [After She Fought for Her Home in the Settlement of Sa Nur, Miryam Adler Produced a Movie About the Race to Give Birth in Religious Society]. *Haaretz*, July 15. https://news.walla.co.il/item/1520378.

Lustick, Ian S. 1988. *For the Land and the Lord: Jewish Fundamentalism in Israel*. New York: Council on Foreign Relations Press.

———. 1999. "Israel as a Non-Arab State: The Political Implications of Mass Immigration of Non-Jews." *Middle East Journal* 53 (3): 417–33.

Lutz, Catherine A., and Lila Abu-Lughod, eds. 1990. *Language and the Politics of Emotion*. Studies in Emotion and Social Interaction. Cambridge: Cambridge University Press; Paris: Editions de la Maison des Sciences de l'Homme.

MacDonald, Mary N., ed. 2003. *Experiences of Place*. Cambridge, MA: Harvard University Press.

Mageni Family. 2003. *Chaim in Judea: In Memory of Chaim Yehuda Mageni*. In Hebrew and English. Kiyrat Arba: Author.

Mahmood, Saba. 2005. *Politics of Piety: The Islamic Revival and the Feminist Subject*. Princeton, NJ: Princeton University Press.

Makdisi, Saree. 2010. "The Architecture of Erasure." *Critical Inquiry* 36 (3): 519–59. doi:10.1086/653411.

Mansour, Camille. 2001. "Israel's Colonial Impasse." *Journal of Palestine Studies* 30 (4): 83–87.

Marotta, Marsha. 2005. "Beyond Maternalism: Women and the Spaces of Political Leadership." *White House Studies* 5 (4): 477–89.

Marty, Martin, and E. R. Scott Appleby, eds. 1991. *Fundamentalisms Observed*. Chicago: University of Chicago Press.

Masalha, Nur. 2007. *The Bible and Zionism: Invented Traditions, Archaeology and Post-Colonialism in Palestine-Israel*. New York: Zed Books.

Massad, Joseph. 2006. *The Persistence of the Palestinian Question: Essays on Zionism and the Palestinians*. New York: Routledge.

Massey, Doreen. 1993. "Power-Geometry and a Progressive Sense of Place." In *Mapping the Futures: Local Cultures, Global Change*, edited by Jon Bird, Barry Curtis, Tim Putnam, and Lisa Tickner, 60–70. New York: Routledge.

———. 1994. *Space, Place, and Gender*. Minneapolis: University of Minnesota Press.
Mattar, Philip. 1992. *The Mufti of Jerusalem: Al-Hajj Amin Al-Husayni and the Palestinian National Movement*. New York: Columbia University Press.
Mazali, Rela. 2002. "In Tow: A Mother's and Daughter's Gendered Departures and Returns." In *Women and the Politics of Military Confrontation: Palestinian and Israeli Gendered Narratives of Dislocation*, edited by Nahla Abdo and Ronit Lentin, 207–33. New York: Berghahn Books.
McKee, Emily. 2015. "Trash Talk: Interpreting Morality and Disorder in Negev/Naqab Landscapes." *Current Anthropology* 56 (5): 733–52.
Memmi, Albert. 1965. *The Colonizer and the Colonized*. Boston: Beacon Press.
Mergui, Raphael, and Philippe Simonnot. 1987. *Israel's Ayatollahs: Meir Kahane and the Far Right in Israel*. London: Saqi.
Mizrachi, Yossef. 1974. "Israeli Sovereignty over the Land of Israel in Light of Jewish Law." In *Kiryat Arba, Hee-Hevron (Kiryat Arba Is Hebron)*. Kiryat Arba: Municipality of Kiryat Arba.
Moors, Annelies. 1995. *Women, Property and Islam: Palestinian Experiences, 1920–1990*. Cambridge: Cambridge University Press.
Moskowitz, Irving. 1998. "A Philanthropist Speaks: Zionism for the 1990s." In *Jews in Places You Never Thought Of*, edited by Karen Primack, 280–81. Hoboken, NJ: KTAV Publishing House.
Neuman, Tamara. 2002. "Alienations of Exilic Return: Russian Immigration and 'Ingathering' in Hebron." In *Realms of Exile: Nomadism, Diasporas, and Eastern European Voices*, edited by Domnica Radulescu, 125–48. Lanham, MD: Lexington Books.
———. 2003. "Transnational Migration and the Religious National Project: The Shifting Contours of Settlement in Hebron." Department of Anthropology and International Studies, Yale University, New Haven, CT, January.
———. 2004a. "The Changing Face of Israeli Settlement: Diaspora, Conversion and the Law of Return." Arlene Schnitzer Judaic Studies Program, Portland State University, Portland, OR, May.
———. 2004b. "In-Gathering and the Pursuit of Religious Sovereignty in the West Bank." In *Transcendent Sovereign: Religious Refashionings Between Diaspora and Empire*. Panel co-organized with Engseng Ho. Society of Cultural Anthropology Conference, Portland, OR, May.
———. 2004c. "Maternal 'Anti-Politics'" in the Formation of Hebron's Jewish Enclave." *Journal of Palestine Studies* 33 (2): 51–70.
———. 2005. "Lost Tribes, Found Nations and Present Pasts in West Bank Settlement." Department of Anthropology, New School University, New York, Februrary.
Neusner, Jacob. 1991. *The Talmud of the Land of Israel*. Vol. 12, *Erubin*. Chicago: University of Chicago Press.
———. 2001. *Understanding Jewish Theology: Classical Issues and Modern Perspectives*. Binghamton, NY: Global Academic Publishing.
———. 2003. "The Religion, Judaism, in America: What Has Happened in Three Hundred Fifty Years?" *American Jewish History* 91 (3): 361–69.
Newman, David. 2005. "From Hitnachalut to Hitnatkut: The Impact of Gush Emunim and the Settlement Movement on Israeli Politics and Society." *Israel Studies* 10 (3): 192–224.
Nora, Pierre. 1989. "Between Memory and History: *Les Lieux de Mémoire*." *Representations* 26 (1): 7–24.
Ochs, Juliana. 2013. *Security and Suspicion: An Ethnography of Everyday Life in Israel*. Philadelphia: University of Pennsylvania Press.

References

Paine, Robert. 1995. "Topophilia, Zionism and 'Certainty': Making a Place Out of the Space That Became Israel Again." In *The Pursuit of Certainty: Religious and Cultural Formulations*, edited by Wendy James, 161–92. New York: Routledge.

Palestinian Central Bureau of Statistics. 2016. "Localities in Hebron Governorate by Type of Locality and Population Estimates, 2007–2016." http://www.pcbs.gov.ps/Portals/_Rainbow/Documents/hebrn.htm.

Pan, Esther. 2005. "Q&A: The Gaza Withdrawal." *New York Times*, August 8. Sec. International. http://www.nytimes.com/cfr/international/slot2_080805.html.

Parfitt, Tudor. 2003. *The Lost Tribes of Israel: The History of a Myth*. London: Phoenix.

Pedahzur, Ami. 2012. *The Triumph of Israel's Radical Right*. New York: Oxford University Press.

Pedahzur, Ami, and Holly McCarthy. 2015. "Against All Odds—the Paradoxical Victory of the West Bank Settlers: Interest Groups and Policy Enforcement." *Israel Affairs* 21 (3): 443–61.

Pedahzur, Ami, and Arie Perliger. 2011. *Jewish Terrorism in Israel*. New York: Columbia University Press.

Peled, Yoav. 2014. *The Challenge of Ethnic Democracy: The State and Minority Groups in Israel, Poland and Northern Ireland*. New York: Routledge.

Pelham, Nicolas. 2009. "Israel's Religious Right and the Peace Process." *Middle East Research and Information Project*. October 12. http://www.merip.org.silk.library.umass.edu/mero/mero101209?ip_login_no_cache=a50aab7e20b3508ecc9e9862233279f7.

Peteet, Julie. 1991. *Gender in Crisis: Women and the Palestinian Resistance Movement*. New York: Columbia University Press.

———. 2009. *Landscape of Hope and Despair: Palestinian Refugee Camps*. Philadelphia: University of Pennsylvania Press.

Peters, Joel, and David Newman, eds. 2013. *Routledge Handbook on the Israeli-Palestinian Conflict*. New York: Routledge.

Portugese, Jacqueline. 1998. *Fertility Policy in Israel: The Politics of Religion, Gender, and Nation*. Westport, CT: Greenwood Publishing Group.

Prentiss, Craig R., ed. 2003. *Religion and the Creation of Race and Ethnicity: An Introduction*. New York: New York University Press.

Primack, Karen. 1998. *Jews in Places You Never Thought Of*. Hoboken, NJ: KTAV Publishing House.

Quandt, William B. 1986. "Camp David and Peacemaking in the Middle East." *Political Science Quarterly* 101 (3): 357–77. doi:10.2307/2151620.

Ram, Uri. 1999. "The Colonization Perspective in Israeli Sociology." In *The Israel/Palestine Question*, edited by Ilan Pappé, 55–80. Hove, UK: Psychology Press.

Rapoport, Tamar, Yoni Garb, and Anat Penso. 1995. "Religious Socialization and Female Subjectivity: Religious-Zionist Adolescent Girls in Israel." *Sociology of Education* 68 (1): 48–61.

Ravitzky, Aviezer. 1996. *Messianism, Zionism, and Jewish Religious Radicalism*. Chicago: University of Chicago Press.

Remennick, Larissa I. 1998. "Identity Quest Among Russian Jews of the 1990s: Before and After Emigration." In *Jewish Survival: The Identity Problem at the Close of the Twentieth Century*, edited by Ernest Krausz and Gitta Tulea, 241–58. New Brunswick, NJ: Transaction Publishers.

———. 2002. "Transnational Community in the Making: Russian-Jewish Immigrants of the 1990s in Israel." *Journal of Ethnic and Migration Studies* 28 (3): 515–30.

Ricoeur, Paul. 1986. *Lectures on Ideology and Utopia*. New York: Columbia University Press.

Riesebrodt, Martin. 1998. *Pious Passion: The Emergence of Modern Fundamentalism in the United States and Iran*. Berkeley: University of California Press.

Robinson, Shira. 2013. *Citizen Strangers: Palestinians and the Birth of Israel's Liberal Settler State*. Stanford, CA: Stanford University Press.

Rose, Jacqueline. 2005. *The Question of Zion*. Princeton, NJ: Princeton University Press.

Rosman, Abraham and Paula G. Rubel. 2015. "'I Love a Parade': Ethnic Identity in the United States and Israel." In *Toward an Anthropology of Nation Building and Unbuilding in Israel*, edited by Fran Markowitz, Stephen Sharot, and Moshe Shokeid, 293–308. Omaha: University of Nebraska Press.

Rutherford, Danilyn. 2016. "Affect Theory and the Empirical." *Annual Review of Anthropology* 45 (1): 285–300.

Ruthven, Malise. 2007. *Fundamentalism: A Very Short Introduction*. Oxford: Oxford University Press.

Said, Edward W. 1992. *The Question of Palestine*. New York: Vintage.

Sanders, Edmund. 2013. "Palestinians in West Bank's Area C Suffer in Limbo." *Los Angeles Times*, May 18. http://articles.latimes.com/2013/may/18/world/la-fg-west-bank-area-20130519.

Saussure, Ferdinand de. 1983[1915]. *Course in General Linguistics*. London: Duckworth.

Savir, Uri. 2010. *The Process: 1,100 Days That Changed the Middle East*. New York: Vintage.

Segev, Tom. 2008. *1967: Israel, the War, and the Year That Transformed the Middle East*. New York: Picador.

Shafir, Gershon. 1989. *Land, Labor and the Origins of the Israeli-Palestinian Conflict, 1882–1914*. Cambridge: Cambridge University Press.

———. 1999. "Zionism and Colonialism: A Comparative Approach." In *The Israel/Palestine Question*, edited by Ilan Pappé, 81–96. Hove, UK: Psychology Press.

Shafir, Gershon, and Yoav Peled. 2002. *Being Israeli: The Dynamics of Multiple Citizenship*. Cambridge: Cambridge University Press.

Shahak, Israel, and Norton Mezvinsky. 2004. *Jewish Fundamentalism in Israel*. New ed. London: Pluto Press.

Shamgar Commission [Israel]. 1994a. Report of the Commission of Inquiry into the Massacre at the Tomb of Patriarchs in Hebron. Excerpts, in English. June 26. https://tinyurl.com/y9klr9yo.

———. 1994b. *Vaʿadat ha-Ḥaḳirah le-ʿInyan ha-Ṭevaḥ bi-Meʿarat ha-Makhpelah, Din ve-Ḥeshbon* [Report of the Commission of Inquiry into the Massacre at Meʿarat ha-Makhpelah].

Sharoni, Simona. 1997. "Motherhood and the Politics of Women's Resistance: Israeli Women Organizing for Peace." In *The Politics of Motherhood: Activist Voices from Left to Right*, edited by Alexis Jetter, Annelise Orleck, and Diana Taylor, 144–60. Hanover, NH: University Press of New England.

Shehadeh, Raja. 1988. *Occupier's Law: Israel and the West Bank*. Rev. ed. Washington, DC: Institute for Palestine Studies.

Shlaim, Avi. 2014. *The Iron Wall: Israel and the Arab World*. 2nd ed. New York: W. W. Norton.

Silverstein, Michael. 1976. "Shifters, Linguistic Categories, and Cultural Description." In *Meaning in Anthropology*, edited by Keith H. Basso and Henry A. Selby, 11–55. Albuquerque: University of New Mexico Press.

Simons, Chaim. 1995. *A Scholarly Study of Dr. Baruch Goldstein's Act in the Cave of Machpelah*. Kiryat Arba: Author.

———. 2003. *Three Years in a Military Compound: Reminiscences of a Hebron Settler*. Kiryat Arba: Author.

Slyomovics, Susan. 1998. *The Object of Memory: Jew and Arab Narrate the Palestinian Village*. Philadelphia: University of Pennsylvania Press.
Smith, Jonathan Z. 1982. *Imagining Religion: From Babylon to Jonestown*. Chicago: University of Chicago Press.
———. 1992. *To Take Place: Toward Theory in Ritual*. Chicago: University of Chicago Press.
Sprinzak, Ehud. 1991. *The Ascendance of Israel's Radical Right*. New York: Oxford University Press.
———. 1999. *Brother Against Brother: Violence and Extremism in Israeli Politics from* Altalena *to the Rabin Assassination*. New York: Free Press.
Sterman, Adiv. 2016. "State Must Recognize Private Orthodox Conversions, Court Says." *Times of Israel*, March 31.
Stern, Moishe. 1997. "Li-Deḥot o Le-Ḳarev?" *Nekuda* 205 (May): 38–42.
Stoler, Ann Laura. 2002. *Carnal Knowledge and Imperial Power: Race and the Intimate in Colonial Rule*. Berkeley: University of California Press.
Strozier, Charles B., David M. Terman, James W. Jones, and Katherine A. Boyd. 2010. *The Fundamentalist Mindset: Psychological Perspectives on Religion, Violence, and History*. New York: Oxford University Press.
Struck, Doug. 1994. "Soldiers Dispute Account of Massacre in Hebron." *Baltimore Sun*, March 18. http://articles.baltimoresun.com/1994-03-18/news/1994077002_1_mosque-galil-arabs.
Stump, Roger W. 2000. *Boundaries of Faith: Geographical Perspectives on Religious Fundamentalism*. Lanham, MD: Rowman & Littlefield.
———. 2008. *The Geography of Religion: Faith, Place, and Space*. Lanham, MD: Rowman & Littlefield.
Tamari, Salim. 2009. *Mountain Against the Sea: Essays on Palestinian Society and Culture*. Berkeley: University of California Press.
Taub, Eliav, and Avi Sasson. 2012. "Geographical and Bureaucratic Aspects Following the Establishment of 'Tombs of Saints.'" *Horizons in Geography*, nos. 81–82: 39–50.
Taub, Gadi. 2011. *The Settlers: And the Struggle over the Meaning of Zionism*. New Haven, CT: Yale University Press.
Thangtungnung, H. 2014. "Origin and Migration of the Zo People." *Man in India* 94 (1–2): 225–40.
———. 2015. "Ethnic History and Identity of the Zo Tribes in North East India." *Journal of North East India Studies* 5 (1): 39–50.
Thrift, Nigel. 2008. *Non-Representational Theory: Space, Politics, Affect*. New York: Routledge.
Todorov, Tzvetan. 1999. *The Conquest of America: The Question of the Other*. Norman: University of Oklahoma Press.
Tuan, Yi-Fu. 1990. *Topophilia: A Study of Environmental Perception, Attitudes, and Values*. New York: Columbia University Press.
Weil, Shalva. 2003. "Dual Conversion Among the Shinlung of North-East India." *Studies of Tribes and Tribals* 1 (1): 43–57.
Weisburd, David. 1989. *Jewish Settler Violence: Deviance as Social Reaction*. University Park: Pennsylvania State University Press.
Weiss, Efrat. 2008. "We'll Go to War over Hebron House, Warn Settlers." *Ynetnews*. http://www.ynetnews.com/articles/0,7340,L-3623990,00.html.
Weiss, Hadas. 2017. "Embedded Politics in a West Bank Settlement." In *Normalizing Occupation: The Politics of Everyday Life in the West Bank Settlements*, edited by Ariel Handel, Marco Allegra, and Erez Maggor, 75–91. Bloomington: Indiana University Press.

Weizman, Eyal. 2012. *Hollow Land: Israel's Architecture of Occupation.* London: Verso.
Wiles, Rich. 2014. "Remembering the Ibrahimi Mosque Massacre." *Al Jazeera,* February 24. http://www.aljazeera.com/indepth/inpictures/2014/02/remembering-ibrahimi-mosque-ma-2014223105915230233.html.
Wolfe, Patrick. 1999. *Settler Colonialism and the Transformation of Anthropology: The Politics and Poetics of an Ethnographic Event.* London: Cassell.
Yiftachel, Oren. 2006. *Ethnocracy: Land and Identity Politics in Israel/Palestine.* Philadelphia: University of Pennsylvania Press.
Yiftachel, Oren, and Avinoam Meir. 1998. *Ethnic Frontiers and Peripheries: Landscapes of Development and Inequality in Israel.* Boulder, CO: Westview Press.
Yiftachel, Oren, and Batya Roded. 2011. "Abraham's Urban Footsteps: Political Geography and Religious Radicalism in Israel/Palestine." In *The Fundamentalist City? Religiosity and the Remaking of Urban Space,* edited by Nezar AlSayyad and Mejgan Massoumi, 177–208. New York: Routledge.
Yuval-Davis, Nira. 1989. "National Reproduction and 'the Demographic Race' in Israel." In *Woman-Nation-State,* edited by Nira Yuval-Davis and Floya Anthias, 92–109. New York: St. Martin's Press.
Zertal, Idith, and Akiva Eldar. 2009. *Lords of the Land: The War over Israel's Settlements in the Occupied Territories, 1967–2007.* New York: Nation Books.
Zorema, J. 2007. *Indirect Rule in Mizoram, 1890–1954.* New Delhi: Mittal Publications.
Zureik, Elia. 2011. "Colonialism, Surveillance, and Population Control: Israel/Palestine. In *Surveillance and Control in Israel/Palestine: Population, Territory and Power,* edited by Elia Zureik, David Lyon, and Yasmeen Abu-Laban, 3–46. New York: Routledge.

Index

Page numbers in italics indicate images.

Abraham, Nabeel, 203n2
Abraham's purchase of Sarah's burial site, 3–4, 29–31, 36, 77, 135, 195n3, 205n19. *See also* Tomb of the Patriarchs (Meʿarat ha-Makhpelah/al-Ḥaram al-Ibrahimi)
Abu Sneinah (Palestinian residential area), 90
Adler, Miryam, 93
Allon, Yigal, 52–53, 72–73
Allon settlement plan, 52–54, 194n4
Amar, Rav Shlomo, 162
Amir, Yigal, 18, 144, 206n26
Amishav (My Nation Returns), 156, 158, 208n7
Aran, Gideon, 54, 84–85, 121, 203n30
Arendt, Hannah, 198n8
Asad, Talal, 12–14, 194n11
Attias, Jean-Christophe, 195n1
Auerbach, Jerold S., 206n24
Avichail, Rav Eliyahu, 153, 155–62, 207n5, 208n7
Avnery, Uri, 61–62
ʿAvodah Zarah (Talmudic tractate), 118–19
Avraham Avinu (Hebron neighborhood), *3*, 90
Ayyash, Yahya, 116, 202n23

Barmash, Pamela, 207n1
Barth, Fredrik, 99
Begin, Menachem, 66
Beit El (settlement), 36
Beit Hadassah, 79–84; Daboya clinic takeover (1979), 72, 79–84, *82*, 88–89; fortified enclave (2005), *81*; original clinic in the wake of 1929 riots, *80*. *See also* Daboya clinic takeover by activist women and children (1979)
Beit HaShalom (Palestinian al Rajabi building), 175–76, 183, 209n22
Beit Jala (Palestinian town), 18, 33, 34–35

Ben Gurion, David, 94
Ben Nun, Rabbi Joel, 172
Ben Shitrit, Lihi, 198n9
Ben-Ari, Eyal, 76–77
Benbassa, Esther, 195n1
Benjamin, Walter, 197n8
Bethlehem, 48–49
Bigelow, Anna, 204n10
Birnbaum, Rav Eliahu, 158
Bnei Akiva movement, 32
Bnei Menashe, 153, 154–66, 177–78, 207n3; animism, 155, 208n6; Avichail's recruitment and conferring of Jewish origins, 153, 155–62, 207n5, 208n7; and biblical tribe of Manasseh, 155, 158, 207n5; Christian background, 155, 160, 208n6; circumcision rituals, 159–60; citizenship status and immigrant absorption, 164–66; contrasts with Ethiopian Jews, 156, 165, 166–67; conversions to Judaism, 162, 163; embrace of ideological settler ethos and Kiryat Arba as home, 163–64; Khaute's views as spokesperson for, 162–64; Kuki-Chin-Mizo ethnic group, 153, 155, 161, 207n3, 207n4; labor and employment, 157; languages, 157, 208n10; as "lost tribe," 153, 154–66, 207n5; population estimates, 155, 207n4; sacrifice rituals, 160; signs and criteria for Jewish origins, 159–62; struggle for acceptance by mainstream Israelis, 163–64; struggle for government recognition as Jews, 162–63
borders/boundaries: cyclone security fencing, 20, 86–87, *106*, 106–7, 115–16; designated by checkpoints, 43, 44–45, 111; ethnic boundaries between Jewish "self" and Palestinian "other," 171; expansion of the *eruv* and the private sphere, 86–87; Green Line,

borders/boundaries (*cont.*)
    4, 7–8, 33, 43, 52–53, 67–68, 182, 193n1, 193n3, 197n5; and ideological uses of religiously-inscribed space, 20, 86–87, 97–99, 106–8, 115–17; immigrant recruitment and intra-Jewish "difference," 151–53, 168–69, 171–72, 173, 177–78; of Kiryat Arba, 20, 86–87, 97–99, *106*, 106–8, 115–17; military policies of separation, 43–44, 196n14; partitioning of the Ibrahimi mosque, 4, 73, 74, 123, 127–28, 132–37, 150, 205n19; remade through bypass roads, 40–41, 44–45, 48–49, 196n10, 196n14, 202n15; settlement along the Jordanian border, 7, 52–54, 194n4; settler routes and expansion of, 97–99, 100–101; settler violence and absence of fixed, 62–63, 197n8
Bowman, Glenn, 204n10
Breaking the Silence (activist group), 41–42
Bridges for Peace (Christian evangelical organization), 188
British Mandate period: land claims during, 58; limits on Jewish immigration and areas of residence in Palestine, 60. *See also* Hebron massacre of 1929
burial sites: Abraham's purchase of Sarah's, 3–4, 29–31, 36, 77, 135, 195n3, 205n19; diaspora and dilemma of, 30–31; Goldstein's gravesite and shrine in Kiryat Arba, 147–50, *148*, 207n31; and Jewish saint worship, 131–32, 205n12; for Jewish victims of 1929 Hebron Massacre, 9, 75; purification issues and graves, 130–32; religious pilgrimage to graves of patriarchs and matriarchs, 131–32, 204–5n11, 205n12; standoff at Hebron's old Jewish cemetery over burial of Nachshon's son, 72, 75–78. *See also* Tomb of the Patriarchs (Meʿarat ha-Makhpelah/al-Ḥaram al-Ibrahimi)
Bus 160 (connecting Jerusalem and Kiryat Arba/Kiryat Arba and Old City of Hebron), 20, 111, 202n15
Butler, Judith, 198n8
bypass roads, 40–41, 44–45, 48–49, 196n10, 196n14, 202n15

Camp David Agreement (1979), 79, 202n21
Canetti, Elias, 200n1
cemetery, Jewish. *See* Jewish cemetery in Hebron
checkpoints, military, 43, 44–45, 111

Christian Peacemaker Teams (CPT), 45
Church of the Holy Sepulcher (Jerusalem), 205n16
circumcision: Muslim ritual of, 159–60; Nachshon's secret circumcisions of her sons at the Tomb of the Patriarchs, 72–74; rituals of the Bnei Menashe, 159–60
Cohen, Anat, 89, 199n12
Cohen, Geula, *82*
colonialism: anticolonial riots during the British Mandate, 4, 72; land confiscation, 7, 45–47, 48, 57, 108; "lost tribes" as colonial trope, 154–55, 207n1, 208n8; Memmi on, 190, 194n8, 206n25; Peteet on, 39; settler colonialism, 6, 7, 39, 194n8; settlers' privileged status and, 206n25; stance of nonrecognition and indifference toward the Palestinian other, 20, 23, 97, 105–13, 209n10
conversions of non-Jews, 151, 158–59, 209n19; the Bnei Menashe, 162, 163; controversies over rabbinic authority to oversee, 172, 209n19; meeting demographic problem of expanding settlements, 209n19; Russians' expedited conversions, 171–72, 177, 209n19; traditional Jewish stance on, 158–59
curfews, 97, 100

Daboya clinic takeover by activist women and children (1979), 72, 79–84, *82*, 88–89; and fortified Beit Hadassah enclave (2005), *81*; and the Gush Emunim, 82–83; Leavitt's settler memoir downplaying, 88–89; left-wing reporter Litani on, 81–83; Nachshon and, 79, *82*; newspaper accounts, 79, 81–83, 198n4; the use of children and the staging of maternalism, 81, 89, 90, 198n5; violence against Palestinian shopkeeper, 83–84
Dalsheim, Joyce, 207n30
Dayan, Moshe, 55, 83, 133–34
de Certeau, Michel, 114
de Leon, Isaac, 119
diaspora, Jewish: dilemma of burial sites and deterritorialized self-definitions, 30–31; settlers' recruitment efforts and "ingathering of exiles," 151–53; settlers' views of, 161–62. *See also* immigrants recruited to Kiryat Arba
Dome of the Rock (Jerusalem), 69, 140, 189
Dominguez, Virginia, 200n3, 200n6

Drobles, Matti, 66–69
Drukman, Haim, 172
Durkheim, Emile, 196n17, 204n8

economic transactions, 108–13; shopping at a Palestinian dry-goods store, 108–10; shopping in Hebron's segregated Old Palestinian market, 110–13, 201–2n14
Eldar, Akiva, 55, 57–58, 127
elections, national: confrontations between settlers and Palestinians over, 116–17, 202n22, 202n23; ideological settlers' influence on, 116, 190–91, 202n22; Kiryat Arba voters, 116–17, 202n22, 202n23; Likud rally in front of Tomb of the Patriarchs, *98*; Likud's rise to power, 66–67
El-Haj, Abu, 208n8
Eliade, Mircea, 204n9
Elon, Amos, 78
Elon Moreh (settlment), 36, 179
Elonei Mamre, 36–37
El-Or, Tamar, 84–85
eruv: definitions, 199n10; expansion of, 86–87; and Sabbath requirements, 86–87; and settlers' carrying of weapons in Kiryat Arba, 86–87
Eshkol, Levi, 53–54
Eshmoret Yitzhak (Kiryat Arba neighborhood), 20
Ethiopian Jews (Beta Israel), 153, 166–69, 177–78; alienation and dissatisfaction with life in Kiryat Arba, 156, 169; contrasts with the Bnei Menashe, 156, 165, 166–67; distinct forms of Jewish observance, 166–67, 168, 208n12; integration and reeducation in Kiryat Arba's absorption center, 153, 156, 165, 166–69; and "lost tribe" role, 153, 156; Sabbath observance, 168; viewed as group in crisis, 168–69, 208n15
ethnicity: Barth on, 99; characteristics that make Jewish settlers identifiable to soldiers, 112–13, 202n17; ethnic boundaries between Jewish "self" and Palestinian "other," 171; formation of ethnic enclaves and ideological uses of, 99, 102–5, 114, 200n2; immigrant recruitment and ethnicizing processes, 151–53, 168–69, 171–72, 173, 177–78; Israel's ethnic character and religious observance of Judaism, 201n7; race vs., 200n2; segregated spaces and entrenchment of ethnic divides, 110–13; settlers' assertion of "difference" between Jews and Palestinians, 39–40; spatial character of ethnic exclusivism, 23, 102–5; Yiftachel's term "ethnocracy" and the state project of territorial expansion, 102–3
Eyal (Kahanist group), 144

Feige, Michael, 124–25, 194n9
fertility rates, 91, 199n16. *See also* reproduction, prolific
fieldwork in Kiryat Arba, 17–21, 186–88; arrival in the settlement, 18–19; ethical dilemmas, 186–88; immediate post-Oslo period, 18; private domestic interactions, 21, 186; public gatherings and public contexts, 186–88; settler violence and handguns, 187–88
Fishbane, Michael, 195n3
Foucault, Michel, 200n1
Frame Analysis (Goffman), 39–40
Franky, Rabbi Rafael, *10*
Free Center, 60
Freund, Michael, 208n7
fundamentalism. *See* Jewish fundamentalism

Gahal (precursor to the Likud), 60
Gaza settlements: and the Bnei Menashe, 155; ideological settlers of, 155, 179–83, *181*; settlers' compensation and resettlement, 180–81, 209n2, 209n20; withdrawal (2005), 24–25, 179–83, 201n9, 209n1; young settlers and clashes, 179
Gazit, Shlomo, 52, 54–57, 63, 197n3, 197n5
Geneva Conventions, 50, 52
Gilo (settlement), 34–35
Givat Harsina (Kiryat Arba neighborhood), 19–20
Goffman, Erving, 39–40, 49
Goldstein, Baruch, 122–24, 138–42, 145–50, 202n23; disregard for secular law and regulations at the Tomb, 138–42; gravesite and shrine, 147–50, *148*, 207n31; and Kach Party, 139–42, 148–49; posthumous remaking as hero and saint, 145–50, 207n32. *See also* Goldstein massacre at the Tomb of the Patriarchs (1994)
Goldstein massacre at the Tomb of the Patriarchs (1994), 122–24, 132–50; and alliance between settlers and military soldiers, 122–24, 132, 205n14; different numbers of Palestinian deaths and injuries, 124, 203n2;

Goldstein massacre (*cont.*)
far-right apologists for, 142–45, 150, 206n22, 206n23; Goldstein's disregard for regulations at the Tomb, 138–42; Goldstein's posthumous remaking as hero and saint, 145–50, 207n32; Kahane and, 139–41, 147, 191, 207n31; Paine on, 201n8; and the partitioned mosque, 127–28, 132–37, 150; and Purim holiday, 122–23, 138, 203–4n6; settlers' complicity and downplaying of violence, 146–47; Shamgar Commission report, 122, 126–27, 132–37, 203n1, 203n3, 204n7
Gonen, Bella, 165–66, 167–69
Green Line (pre-1967 border or 1949 Armistice lines), 4, 7–8, 33, 43, 52–53, 67–68, 182, 193n1, 193n3, 197n5
guns and weapons in settlements: armed settlers, 24, 51, 86–87, 96–97, 142, 143, 187–88; and fieldwork dilemmas, 187–88; during Sabbath worship, 86–87. *See also* violence, settler
Gush Emunim (Bloc of the Faithful), 15, 58–59, 66, 69, 198n7; Aran and belief system of, 203n30; Avichail and, 157; protesting dismantling of Yamit settlement in the Sinai, 182, 202n21, 209n21; and women's takeover of Daboya clinic (Beit Hadassah), 82–83
Gush Etzion (settlement), 32–33, 43, 44

Haaretz: Elon on Nachshon's activism at Old Jewish cemetery, 78; Hass on Tomb of the Patriarchs at twenty-first anniversary of Goldstein massacre, 207n34; Litani on Daboya clinic takeover, 81–83
Haetzni, Elyakim, 140–42
halakha (Jewish law): and Abraham's purchase of Sarah's burial site, 195n3; division of laws to be implemented in human time/messianic era, 15–16, 117–18, 120; kashrut (dietary law), 65; Mizrachi on Noahide laws and category of *ger toshav* (resident alien), 120, 203n29; Mizrachi's justifications to legitimate settlement and territorial expansion, 117–20; rabbinic commentaries on land and territory in, 117–20; remaking as place-dependent, 11
Halhoul (Palestinian city), 18, 33, 36–40, 196n8
Hall of Abraham and Sarah (Tomb of the Patriarchs), 135

Hall of Isaac and Rebecca (Tomb of the Patriarchs), 123, 128, *129*, 135, 137, 138, 205n18
Hamas, 142–43, 202n23
Hammer, Zevulun, 83
Ha-ʿOlam ha-Zeh (news weekly), 61
Ha-ʿOlam ha-Zeh-Koaḥ Hadash, 61–62
al-Ḥaram al-Ibrahimi (Abraham's Sanctuary), 3. *See also* Tomb of the Patriarchs (Meʿarat ha-Makhpelah/al-Ḥaram al-Ibrahimi)
Haredi Judaism, 85, 172
Harel, Asaf, 195n12
Harsina (Mount Sinai), 45–46
Harvey, David, 194n6
Hass, Amira, 207n34
Hassan, Rabbi, *10*
Hassner, Ron, 130, 204n9
Hayden, Robert, 204n10
Hebrew Scriptures and Jewish canonical texts, reinterpreting, 28–31, 33, 36–37; Abraham's purchase of Sarah's burial site, 3–4, 29–31, 36, 77, 135, 195n3, 205n19; to authorize renewal of the Old Jewish cemetery, 77–78; Avichail on the Bnei Menashe as "lost tribe," 157–58; biblical passages that forbid conquest, 118–19; to confer contemporary property rights, 28–31, 36–37; to defend land confiscation, 47–48; and question of Bible's historicity, 31, 33, 37, 196n6; symbolic biblical violence, 203–4n6, 207n31; and tribe of Manasseh, 155, 158, 207n5
Hebron, Palestinian Old City of, 2–4; Bus 160's route, 20, 111, 202n15; curfews, 97, 100; Daboya clinic (Beit Hadassah), 72, 79–84, *80*, *81*, *82*, 88–89; Hebron massacre (1929), 4, 9, *10*, 59, 72, 75, 80–81, 205n17; Israeli Independence Day celebrations in, 96, 99–102, 200n3, 200–201n6; "Jewish Quarter," *3*, 3–4, 80, 177, 209n23; map, *3*; Old Jewish cemetery, 72, 75–78; Palestinian Park Hotel takeover (1968), 50, 54–56; realms of authority and Hebron's H1 and H2 zones, *3*, 4, 41; restrictions on Palestinian movement, 3, 4, 40–41, 42–45, 97, 100, 111, 113–14; security regime, 42–45; settler tours, 22, 32–41, 47–48, 195n4. *See also* ideological settlers' initial tactics and transgressions in occupied Hebron; Kiryat Arba; maternalist activism and property takeover in Old City of Hebron; Tomb of the

Patriarchs (Meʿarat ha-Makhpelah/al-Ḥaram al-Ibrahimi)
Hebron Hills, 4
Hebron massacre of 1929, 4, 9, *10*, 30, 33, 59, 72, 75, 80–81, 88–89, 205n17; Leavitt's memoir, 88–89; and the Old Jewish cemetery, 75; reparations issue, 61, 198n2
Hebron Protocol (1997), 4
Heimat model of national belonging, 16
Herut Party, 66, 67
Herzfeld, Michael, 201n10
Herzog, Chaim, 205n15
hesder yeshiva, 64–65, 117, 175, 203n26
"hilltop youth," 25, 183, 190
hitnaḥalut, 193n3
housing: demolition of Palestinian, 42, 45–46, 61; Kiryat Arba, 18–20, *19*, 42, *51*; subsidized settler, 170, 176–77

ideological formation, religion as, 4–6, 184–85
ideological settlers (terms and definitions), 6–7, 25, 193n1, 193n3, 194n5; *hitnaḥalut*, 193n3; *mitnaḥalim*, 68, 193n3; and Modern Orthodoxy, viii; and phrase "our public," 206n28. *See also* Jewish settlers, types of
ideological settlers' initial tactics and transgressions in occupied Hebron, 6–8, 22–23, 50–69; adaptation of Jewish observance to military context, 22, 63–64; building a *mikve* within military headquarters, 63; early violence against Palestinians, 62–63, 69; exploitation of legal gray zone of the military occupation, 7–8, 22, 51–52, 55, 56–58, 108, 138, 187; Gazit and, 52, 54–56, 197n3; illegal entrance into Hebron, 22, 50, 54–56; illegality, 4, 22, 50, 54–64; implications for democracy and rule of law, 61, 62; "kiosk incident" (1968) and Knesset debate, 58–64; left-wing Knesset members' arguments, 61–62; Likud Party and official approval of settlements, 66–69; military accommodation of, 50, 63–64; Palestinian Park Hotel takeover (1968), 50, 54–56; request for a voting booth, 64; request for extension of municipal services, 64–65; right-wing Knesset supporters of, 59–61; separate taxi service, 64; Simons's memoir on first three years in Hebron, 64–66; spread of ideas through religious education and first yeshiva, 64–65; temporary relocation to Hebron's military headquarters, 42, 50–51, 58–59, 63; use of media and publicity, 64
immigrants recruited to Kiryat Arba, 24, 151–78; the Bnei Menashe of India, 153, 154–66, 177–78, 207n3; conversions to Judaism, 151, 158–59, 162, 163, 171–72, 177, 209n19; dissatisfaction with life in Kiryat Arba, 156, 169; economic migrants, 170, 176, 208n16; embracing settlement ethos and becoming observant ideological settlers, 163–64, 174–76; Ethiopian Jews (Beta Israel), 153, 166–69, 177–78; ethnicizing processes and intra-Jewish "difference," 151–53, 168–69, 171–72, 173, 177–78; immigration law and citizenship, 164–66, 170–71; Jewish diaspora and "ingathering of exiles," 151–53; and Kiryat Arba's absorption center, 153, 156, 165, 166–69; labor and employment, 157, 176–77; language issues and Hebrew fluency, 170, 171, 172–73; "lost tribes," 153–66, 207n1, 208n8; non-Jewish Russians, 171–72, 175, 209n18, 209n19; and Palestinians, 154, 164, 175–76; resistance to ideological settler ethos, 170, 208n17; Russians, 153–54, 169–77; secular-oriented Russians, 170, 173, 176–77; and settlers' emphasis on Jewish origins, 153, 155–62, 178, 208n8; and Zionist immigrant-settler project, 94, 208n8
Inbari, Motti, 124–25
India, immigrants from, 153, 155; Manipur region, 153, 155, 162–64; Mizoram region, 153, 155. *See also* Bnei Menashe
infrastructure: bus routes, 20, 111, 202n15; bypass roads, 40–41, 44–45, 48–49, 196n10, 196n14, 202n15; demolitions, 42, 45–46, 61; military checkpoints, 43, 44–45, 111; municipal services for settlers, 64; roadblocks, 207n34n44; separation barrier, 43; subsidized housing for settlers, 170, 176–77; tunnels, 43, 48–49
Intifada: First, 33, 101, 205n18; Second, 44
Islam: Asad on radical/fundamentalist religion and, 12–13; Islamic character of al-Ḥaram al-Ibrahimi (Tomb of the Patriarchs), 1, 127, 128, 193n2, 205n13; Jewish-Muslim aural competition at al-Ḥaram al-Ibrahimi (Tomb of the Patriarchs), 132; Jewish-Muslim battle for control of al-Ḥaram al-Ibrahimi (Tomb of the Patriarchs), 3–4, 29–30, 128–29, 132–37, 204n8; partitioning

Islam (*cont.*)
  of the Ibrahimi mosque (Tomb of the Patriarchs), 4, 73, 74, 123, 127–28, 132–37, 150, 205n19; religious holidays, 137; ritual of circumcision, 159–60; Temple Movements and ideological settlers' efforts to worship at Islamic sites, 189–90
Israeli Central Command, 50
Israeli Defense Forces (IDF): and Nahal settlements, 53; standoff with women settlers at Hebron's Old Jewish cemetery (1975), 75–78
Israeli General Security Service (GSS) (Shabak), 144, 206n26
Israeli Independence Day celebrations in Hebron, 96, 99–102, 200n3, 200–201n6; appropriation of state symbols for religious uses, 101–2, 200–201n6; flag displays and fireworks, 100–101; the open square facing Meʿarat ha-Makhpelah, 100; prayer at the Tomb of the Patriarchs, 102
Israeli Ministry of Foreign Affairs, 90
"Israeli Sovereignty over the Land of Israel in Light of Jewish Law" (Mizrachi), 117
Israeli state/nation: critics of ideological settlers, 184; demographic emphasis and pronatalism, 94–95; "ethnocracy" and state project of territorial expansion, 102–3; Gaza withdrawal (2005) and reexamination of settler-oriented Judaism, 181–83; ideological settlers' growing influence beyond Hebron, 188–91; Independence Day celebrations, 96, 99–102, 200n3, 200–201n6; Kiryat Arba voters in 1996 national election, 116–17, 202n22, 202n23; legal gray zone of the military occupation, 7–8, 22, 51–52, 55, 56–58, 108, 138, 187; settlements and the nationalist project, 6, 16–17, 181–83; and settler violence, 149; settlers' distrust of the state and government surveillance, 21, 144–45
Issacharoff, Avi, 207n31

al-Jaʾabari, Muhammad Ali, 62
Jaber, Atta, 45–47
Jaber, Habah, 45–47
Jabotinsky, Zeev, 67–68
Jerusalem Temple, 131, 204–5n11
Jewish Agency, Settlement Department of, 67
Jewish cemetery in Hebron: inscriptions of 1929 massacre victims' names, 75; settlers' intention to establish, 75–76; standoff between IDF and women settlers over burial of Nachshon's son, 72, 75–78
Jewish fundamentalism: Asad on religious traditions and category of "fundamentalism," 12–13; immigrant settlers' embrace of, 163–64, 174–76; and messianism, 14–15; and religion as ideological formation, 4–6, 184–85; traditional scholarship on, 14–15; women in fundamentalist communities, 71, 84–85
Jewish mysticism, 47–48
Jewish observance, ideological settlers': adaptation to military context in occupied Hebron, 22, 63–64; building a *mikve* within military headquarters, 63; burial, 30–31, 130–32; circumcision, 72–74, 159–60; conversions of non-Jews, 151, 158–59, 209n19; the *eruv*, 86–87, 199n10; of Ethiopian Jews (Beta Israel), 166–67, 168, 208n12; exilic ingathering, 24, 151–53, 208n8; finding lost tribes, 153–66, 208n8; gendered aspects, 70–71, 85, 91; and Israel's ethnic character, 201n7; kashrut, 64; Likud Party and, 66; lived religion, 15–16; *masorti* (traditionally observant), 19; messianism, 14–16, 27–28, 121; military accommodation of, 63–64; Passover, 22, 160; pilgrimage to graves of patriarchs and matriarchs, 131–32, 204–5n11, 205n12; Purim, 122–23, 138, 185, 203–4n6; rabbinical authority, 172, 209n19; religion as ideological formation, 4–6, 184–85; religious judges (*dayanim*), 162–63; at remade biblical sites, 5; Sabbath, 86–87, 96–99, 102, 107, 168, 199n10; saint worship, 131–32, 205n12; shifts in, 5–6. *See also* halakha (Jewish law)
"Jewish Quarter" in Old City of Hebron, 3, 3–4, 80, 177, 209n23
Jewish settlers, types of, 6–8, 25, 193n1, 193n3, 194n5; agricultural, 68, 94, 197n11, 209n21; "hilltop youth," 25, 183, 190; ideological, 6–7, 25, 193n1, 193n3, 194n5, 197n11; military settlers in Nahal, 53–54; *mitnaḥalim*, 68, 193n3; pre-state Zionist, 39, 68, 94, 108, 197n11, 208n8; quality-of-life, 4–5, 34–35, 176, 193n3
Jewish Underground (ha-Maḥteret ha-Yehudit), 69, 140, 145, 199n12
Josephus, 38

Index

Judea, 12, 35, 199n16
June 1967 war, 4, 80, 194n4

Kach Party, 69, 139–42, 206n21; Goldstein and, 139–42, 148–49; Kahane and, 69, 139–40, 206n21; and term *Kach*, 198n12
Kahane, Meir, 191, 206n21; and Goldstein massacre, 139–41, 147, 191, 207n31; Haetzni on, 140–41; Kach Party, 69, 139–40, 206n21; Kahanists in Kiryat Arba, 69, 144; Knesset supporters, 191
Karp, Yehudit, 73
Karp Commission investigation (1984), 73–74, 84
kashrut, settler remaking of, 64
Kastel, Rabbi Meir Shmuel, *10*
Katzover, Benny, 58–59
Kawasmeh, Fahd, 50, 197n1
Kelly, Tobias, 62
Kfar Etzion, 65
Khaute, Tzvi, 162–64
Kimmerling, Baruch, 108, 201n13
"kiosk incident" (1968), 58–64; Knesset debate, 59–63; left-wing arguments of Avnery, 61–62; right-wing supporters against the expulsion of the "kiosk four," 59, 60; right-wing supporters' arguments for continuity of Jewish presence in Hebron and collective restitution of property, 59–61
Kiryat Arba, 2–4, *3*, 17–21, *19*, *51*; archival photo (1971), *51*; atmosphere of small-town suspicion and paranoia, 20–21; Baruch Goldstein's burial site, 147–50, *148*, 207n31; Bnei Menashe members' embrace of life in, 163–64; boundaries and borders, 20, 86–87, 97–99, *106*, 106–8, 115–17; Bus 160's route to, 20, 111, 202n15; conducting fieldwork in, 17–21, 186–88; cyclone security fencing, 20, 86–87, *106*, 106–7, 115–16; environmental awareness and water use, 107, 201n11; eruv for Sabbath observance, 86–87; Ethiopian Jews' dissatisfaction with life in, 156, 169; founding, 4, 33; gendered dimensions of daily life, 85; *hesder* yeshiva of, 64–65, 117, 175, 203n26; immigrant absorption center, 153, 156, 165, 166–69; map, *3*; physical layout and residential neighborhoods, *19*, 19–20; public and private spheres, 86; Purim celebration in, 122–23, 185; soldiers' security patrols, 44–45, 86–87, 122–23; summer fires and blurred boundaries, 115–16; voters in 1996 national election, 116–17, 202n22, 202n23. *See also* immigrants recruited to Kiryat Arba
Kiryat Gat, 179–80
Knesset: Kahane supporters, 191; "kiosk incident" debate (1968), 59–63; and women's takeover of Daboya clinic, 83
Kook, Avraham Itzhak HaCohen, 15
Kook, Zvi Yehuda, 15, 50, 156
Kravel-Tovi, Michal, 209n19
Kuki-Chin-Mizo ethnic group, 153, 155, 161, 207n3, 207n4. *See also* Bnei Menashe

land: confiscation of Palestinian, 45–47, 48, 57, 108; cultivation and ownership, 48, 57, 108; invocations of Jewish mystical tradition to defend confiscation, 47–48; land claims during British Mandate period, 58; Ottoman land laws and ownership, 48, 57, 94, 108, 201n13; settlers' attachments to, 47, 196–97n17, 207n30; Zionist pre-state settlements and land ownership, 39, 94, 108, 201n13
Land of Israel (theological concept), 47, 103, 117–18, 196n9, 207n30
Landau, Haim, 83
law, Jewish. *See* halakha (Jewish law)
law/legality: democracy and international, 8; and Goldstein massacre, 138–42; Haetzni's call to disobey, 141–42; ideological settlers and illegality, 4, 22, 50, 54–64; ideological settlers' illegal entrance into Hebron, 22, 50, 54–56; immigration law and citizenship, 164–66, 170–71; inherited legal codes, 57; Karp Commission's investigations of Nachshon and law enforcement, 73–74, 84; Knesset debate on settler transgressions and implications for democracy and rule of law, 61, 62; legal gray zone of the occupation, 7–8, 22, 51–52, 55, 56–58, 108, 138, 187; Ottoman land laws, 48, 57, 94, 108, 201n13; regulations at the Tomb of the Patriarchs, 137–42; settlers' disregard for, 137–42
Leavitt, June, 87–89
Lehava, 206n21
Levinger, Miriam, *82*
Levinger, Rabbi Moshe, 54, 55
Levinson, Chaim, 207n31
"Li-Deḥot o Le-Ḳarev?" (Stern), 172

Likud Party: Drobles and settlements, 67–69; election rally in front of Tomb of the Patriarchs, *98*; Netanyahu, *98*, 116, 191; official approval of settlements, 66–69; position on Palestinians, 68; post-Oslo, 18; rise to power, 66–67; and settlers' religious observance, 66; Sharon and support for religious settlement, 66–67, 83, 197n9, 201n9; supporters in Kiryat Arba, 190–91, 202n22; tactical alliance with settlers, 67–69
Lior, Rav Dov, 66, 140
Litani, Yehuda, 81–83
locality, 16–17; as bridge to participation in the national project, 16–17; burial sites and diasporic self-definitions, 30–31; concentrating diverse people and pasts into a settled locale, 152; *Heimat* as mode of national belonging, 16; place-based attachments to specific sites, 8, 11–12, 17, 194n8; religion and, 12, 810
"lost tribes," 153–66, 207n1, 208n8; American frontier homesteaders, 154–55; biblical tribe of Manasseh, 155, 158, 207n5; the Bnei Menashe, 153, 154–66, 207n5; Ethiopian Jews (Beta Israel), 153, 156; and immigration law/citizenship, 164–66. *See also* immigrants recruited to Kiryat Arba
Lustick, Ian, 124–25

Mageni, Chaim, 32–41, 47–48
Maimonides (Rambam), 118–19, 120
Majal Shams (Druze village in the Golan Heights), 107–8
Manasseh, lost tribe of, 155, 158, 207n5
Manipur region of India, 153, 155, 162–64
Mar Elias, 204n10
masorti (traditionally observant), 19
Massey, Doreen, 194n6
maternalist activism and property takeover in Old City of Hebron, 23, 70–95; children's deaths as rationales for permanent habitation, 77–78, 90; children's use in, 72–74, 77–78, 81, 88–90, 198n5; circumcisions of Nachshon's sons at the Tomb of the Patriarchs, 72–74; Daboya clinic takeover, 72, 79–84, *81*, *82*, 88–89; emphasizing children's vulnerability and victimization, 88–90; expansion of the *eruv*, 86–87; and gendered aspects of Jewish observance, 70–71, 85, 91; Karp Commission's investigations of law enforcement and, 73–74, 84; Leavitt's memoir focusing on motherhood, 87–89; left-wing journalists on, 78, 81–83; maternalist justifications for Goldstein massacre, 146–47; Nachshon's invocation of the Talmud, 77–78; Nachshon's roles, 72–79, *82*; relation to traditional gender roles and domesticity, 80–85, 198n8, 198n9; and settler violence, 84; Shalhevet Pas's death and funeral, 90; staging maternal grief and sacrifice, 76–78; standoff at Jewish cemetery over Nachshon's infant son's burial, 72, 75–78; upholding patriarchal values, 84–85; and women's prolific reproduction, 71, 85, 89, 91–93, 94, 199n18, 209n19
Meʿarat ha-Makhpelah (Multiple Cave), 3. *See also* Tomb of the Patriarchs (Meʿarat ha-Makhpelah/al-Ḥaram al-Ibrahimi)
Memmi, Albert, 190, 194n8, 206n25
Meretz Party, 149
Merkaz ha-Rav Yeshiva (Jerusalem), 32, 156, 157
messianism: halakhic division of laws to be implemented in human time/messianic era, 15–16, 117–18, 120; ideological settlers and messianic ideals, 14–16, 27–28, 121; messianic redemption, 15, 195n12; scholarship on Jewish fundamentalism and, 14–15; and women's prolific reproduction, 92, 199n18
military, Israeli: checkpoints, 43, 44–45, 111; conscientious objectors, 22, 41–45, 196n12; and first settlers' illegal entrance into Hebron, 22, 50, 54–56; and first settlers' religious observance, 63–64; and Goldstein massacre at Tomb of the Patriarchs, 122–24, 132, 205n14; how Jewish settlers are identifiable to soldiers, 112–13, 202n17; ideological settlers' service in, 196n12, 200n4; partitioning of the Ibrahimi Mosque under British Mandate, 73, 74; policies of separation between settlers and Palestinians, 44, 196n14; presence at Tomb of the Patriarchs, 122–24, 132, 134–35, 137, 205n14, 207n34; relationships between settlers and, 42, 122–24, 132, 196n12, 200n4; security patrols of Kiryat Arba, 44–45, 86–87, 122–23; and settlers' forays into Palestinian space, 96–97, 100–101, 200n4; West Bank security regime, 42–45
mitnaḥel (settler beyond the Green Line), 68, 193n3
Mizoram region of India, 153, 155

Mizrachi, Simha, 172–74
Mizrachi, Yosef, 117–20; on acceptance of Noahide laws and category of *ger toshav*, 120, 203n29; justifications for settlement in Jewish law and rabbinic commentaries, 117–20; on land as religious inheritance, 117; reading of present concerns into the past, 119–20
Mizrachi Jews, 179
Modern Orthodoxy, viii. *See also* ideological settlers (terms and definitions)
Moledet (right-wing party), 116, 202n22
Moskowitz, Irving, 156
motherhood. *See* maternalist activism and property takeover in Old City of Hebron

Nachmanides (Ramban), 119–20
Nachshon, Sarah, 72–79, *82*; and Daboya Clinic takeover, 79, *82*; Elon on fanaticism of, 78; invocation of the Talmud, 77–78; secret circumcision of her sons at Tomb of the Patriarchs, 72–74; staging of maternal grief and sacrifice, 76–77; standoff at Old Jewish cemetery over burial of infant son, 72, 75–78. *See also* maternalist activism and property takeover in Old City of Hebron
Nahal settlements, 53–54
Nakba (Palestinian expulsion in 1948), 6
Narkiss, Uzi, 50
National Religious Party (NRP), 59, 198n7, 202n22
nationalism, 2, 9, 11–12, 16–17, 33, 35, 37, 38; "ethnocracy" and Israeli project of territorial expansion, 102–3; as ethnonationalism, 11–12, 66–67, 101, 103; Israeli Independence Day celebrations in Hebron, 96, 99–102, 200n3, 200–201n6; in Israeli vs. Palestinian claims to the Tomb as heritage, 91; Likud's, 66–67; and pronatalism, 94–95; Russian immigrants' transnational orientation, 170, 208n17; settlements and the Israeli national project, 6–7, 16–17, 181–83; suppression of Palestinian, 66; trope of sacrifice in, 77; Zionist demographic emphasis and, 94, 208n8; Zionist secular nationalist project, 118, 203n28
Nativ, Zehava, 146–47
Nekudah (religious settler journal), 172
Netanyahu, Benjamin, *98*, 116, 191
Neusner, Jacob, 131, 204–5n11
New Communist List, 63

New Kach 2001–2003, 206n21
nonrecognition and indifference toward Palestinians, stance of, 20, 23, 97, 105–13, 209n10; antecedents in frontierism of pre-state settlement, 108; in economic transactions, 108–13
Nora, Pierre, 11
Northern Ireland, sectarianism and Orange parades in, 200n2

Occupation of the West Bank, 6–8; and Breaking the Silence, 41–42; building permits, 45, 67, 174; bypass roads, 40–41, 44–45, 48–49, 196n10, 196n14, 202n15; curfews, 97, 100; early settler violence, 62–63, 69; first ideological settlers' tactics in occupied Hebron, 6–8, 22–23, 50–69; Hebron's military headquarters, 42, 50–51, 58–59, 63; legal gray zone of, 7–8, 22, 51–52, 55, 56–58, 108, 138, 187; military checkpoints, 43, 44–45, 111; military security patrols, 44–45, 86–87, 122–23; Palestinian land confiscations, 45–47, 48, 57, 108; restrictions on Palestinian movement, 3, 4, 40–41, 42–45, 97, 100, 111, 113–14; security regime, 42–45. *See also* infrastructure
origins, ideological settlers' emphasis on, 1–2, 28, 74; attachments to land, 47, 196–97n17, 207n30; the Bnei Menashe and signs/criteria for Jewish origins, 159–62; and immigrant-settler Zionist project, 94, 208n8; interpretive strategies emphasizing continuous Jewish presence in Hebron, 30, 34–35, 196n6; "lost tribes" and immigrant recruitment, 153–66, 178, 207n5, 208n8; place-based emphases, 8–12, 27–28, 103–4, 194n6, 194n9, 194n10; recasting Arabic place names, 34–35; religion and locality, 8, 9
Oslo Accords (1993), 4, 17–18; collapse of, 17–18; Haetzni and calls to disregard secular law, 141–42; Palestinian suicide bombings and, 18, 44; Rabin assassination in the wake of, 144–45; and settler violence, 124–25
others/othering, 49, 99; erasure, 32, 36, 39–40; ethnic boundaries between Jewish "self" and Palestinian "other," 171; ethnic exclusivism, 23, 102–5; and religious violence, 127; segregated spaces, 110–13; settlers'

others/othering (*cont.*)
assertion of "difference" between Jews and Palestinians, 39–40; stance of nonrecognition and indifference toward Palestinians, 20, 23, 97, 105–13, 209n10. *See also* ethnicity
Ottoman land laws, 48, 57, 94, 108, 201n13

Paine, Robert, 103–4, 201n8, 207n30
Palestinian Authority, 4, 18, 33, 40–41, 45
Palestinian Liberation Organization (PLO), 17, 176, 197n1
Palestinian Park Hotel takeover (1968), 50, 54–56
Palestinian refugees, 17, 108, 198n2; al-Arroub camp, 38; Deheishe camp, 18; Gaza, 180–82; Gazit on occupation and, 55–56; Peteet on early Zionists and, 39; UNRWA, 80
Palestinians: Bethlehem, 48–49; and checkpoints, 43, 44–45, 111; confrontations with settlers, 46, 83–84, 115–16, 132, 205n18; curfews, 97, 100; drivers and identity cards, 113–14; Drobles and Likud's position on, 68; farmers living adjacent to settlements, 22, 45–46; and Gaza withdrawal (2005), 180–81; Goldstein massacre casualties, 124, 203n2; Knesset member Toubi on settlers' claim to Hebron, 63; land confiscations, 45–47, 48, 57, 108; Likud position on integration vs. transfer of, 68; marginalization through recruitment of immigrants to Kiryat Arba, 154, 164, 175–76; military policies of separation between settlers and, 44, 196n14; military strategy of instilling fear among, 44; restrictions on movement, 3, 4, 40–41, 42–45, 97, 100, 111, 113–14; settler incursions into areas/property, 96–102, *98*, 107, 116, 190, 200n5; settlers' assertion of ethnic "difference" between Jews and, 39–40; settlers' harassment of, 45–46; settlers' stance of nonrecognition and indifference toward, 20, 23, 97, 105–13, 209n10; shopkeepers, 83–84, 108–10; town of Halhoul, 18, 33, 36–40, 196n8; viewed as foreign population, 57, 197n5; Zionist views of, 39. *See also* Palestinian refugees
Pas, Shalhevet, 90
Pas, Yitzhak, 90

Passover: Avichail on sacrifice rituals of the Bnei Menashe, 160; and settlers' illegal entrance into Hebron (1968), 22, 50, 55–56
Pathans of Afghanistan and Pakistan, 158, 161
Peres, Shimon, 76, 116–17, 135, 202n24, 203n25
Peteet, Julie, 39
place-based emphases. *See* locality; origins, ideological settlers' emphasis on; spatial practices and ideological uses of space
pluralism. *See* religious pluralism
Popular Front for the Liberation of Palestine (PFLP), 202n22
population: creating and distributing a national Jewish majority, 6; demographic battles, 89, 94; displacing native populations in settler colonialism, 7, 180; fertility rates, 91, 199n16; immigrant recruitment to Kiryat Arba, 154, 164, 175–76; Kiryat Arba's diverse, 24; prolific reproduction and, 71, 85, 89, 91–93, 94, 199n18, 209n19; Zionist demographic emphasis and, 94, 208n8
Porat, Rabbi Hanan, 54
Poraz, Avraham, 162–63
power: Asad's approaches to the social and power-laden dimensions of religion, 12–13, 194–95n11; asymmetrical structures, 5, 7, 103, 115, 149–50, 185, 190, 206n25; as inscribed in process of expanding settlement boundaries, 103, 115; religious settler women remaking existing relationships of, 83, 85; religious space as medium of, 8–9, 14; and settler violence, 146, 149–50; settlers' increasing influence and, 58, 183, 185; spatiality and, 13–14, 96, 194–95n11, 200n1; the state's, 105
Purim: biblical violence, 203–4n6; and Goldstein massacre at Tomb of the Patriarchs, 122–23, 138, 203–4n6; in Kiryat Arba, 122–23, 185

rabbinic commentaries on Jewish law, 117–20; analogical reasoning, 119–20; and Mizrachi's discussions of land and territory to legitimate settlement, 117–20; Nachmanides (Ramban), 119–20; Rambam (Maimonides), 118–19, 120; on settlement and conquest, 118–19
Rabin, Yitzhak, 18, 143–45, 147, 206n26

Rachelim outpost near Shilo, 84–85
Rambam (Maimonides), 118–19, 120
Ramban (Nachmanides), 119–20
Raphael, Yitzhak, 59–60
Rashi, 36, 119–20
Ravitsky, Aviezer, 54
Raviv, Avishai, 144, 206n26
Raziel-Naor, Esther, 60
religious certainty, 103–4
religious observance. *See* Jewish observance, ideological settlers'
religious pilgrimage: graves of patriarchs and matriarchs, 131–32, 204–5n11, 205n12; Jerusalem Temple, 131, 204–5n11
religious pluralism: and settler violence, 132, 205n15; and the Tomb of Patriarchs, 132, 137, 205n15
religious tradition and social change: Asad on religious traditions and category of "fundamentalism," 12–13; settlers' interpretations of, 12–16
religious violence. *See* violence, settler
Remennick, Larissa I., 208n17, 209n18
reproduction, prolific, 71, 85, 89, 91–93, 94, 199n18, 209n19; and messianic ideals, 92, 199n18; settlers' fertility rates, 91, 199n16. *See also* population
Russian immigrants in Kiryat Arba, 153–54, 169–77; ambiguous religious identity, 170–72, 175; citizenship and immigration law, 170–71; economic reasons for moving to settlements, 170, 176, 208n16; embrace of settlement ethos and Jewish fundamentalism, 174–76; employment, 176–77; expedited conversions, 171–72, 177, 209n19; government compensation to leave settlements, 173–74, 209n20; Hebrew language and fluency issues, 170, 171, 172–73; long-time settler's views of, 170, 172–75, 177, 208n17; non-Jewish, 171–72, 175, 209n18, 209n19; secular-oriented, 170, 173, 176–77; settlement housing, 170, 176–77; transnational orientation and resistance to ideological settler ethos, 170, 208n17

Sabbath observance: *eruv* and its meanings, 199n10; by Ethiopian Jews (Beta Israel) in Kiryat Arba, 168; expansion of the *eruv* and boundary of the private sphere, 86–87; as model for Israeli Independence Day celebrations in Kiryat Arba, 102; requirement regarding carrying in public, 86–87; requirement to stay put, 86; settler crowds' walking to/from Me'arat ha-Makhpelah, 96–99, *98*, 107
Saladin (Salah ad-Din al-Ayyubi), 205n13
Samaria, 12, 35, 199n16
settlement policies: Allon plan, 52–54, 194n4; Drobles and, 67–69; Likud, 66–69, 83, 197n9, 201n9; Sharon and, 66–67, 83, 197n9, 201n9
settlements, definitions of, 6–7, 25, 193n1, 193n3, 194n5
settlements, dismantling of: Gaza withdrawal (2005), 24–25, 179–83, 201n9, 209n1; Hebron, 183–84; as *nesigah/hitnatkut*, 179–80; and reexamination of settler-oriented Judaism, 181–83; settlers' compensation and resettlement, 174, 180–81, 209n2, 209n20; Yamit in the Sinai (1982), 182, 202n21, 209n21
settler colonialism, 6, 7, 39, 194n8
settler tours of Hebron, 22, 32–41, 47–48, 195n4; claiming Halhoul as a Jewish site, 36–40, 196n8; in contrast with the critical accounts of a former soldier, 22, 41–45, 196n12; emphasizing Jewish origins and continuous presence in Hebron, 34–35, 196n6; invocation of Jewish mystical tradition, 47–48; invoking the Bible to justify settlement, 33, 36–37; recasting Arabic place names, 34–35; reference to a site of sacrifice (a missing altar), 37–38
settler violence. *See* violence, settler
settlers. *See* ideological settlers (terms and definitions); Jewish settlers, types of
Shabak. *See* Israeli General Security Service (GSS) (Shabak)
Shafir, Gershon, 94
Shamgar, Meir, 56–57, 205n1
Shamgar Commission report on Goldstein massacre (1994), 122, 126–27, 132–37, 203n1, 203n3, 204n7
Sharon, Ariel, 66–67, 83, 104–5, 197n9, 201n9
Shaul, Yehuda, 41–45, 196n12
Shavei Israel, 158, 208n7
Shinlung, 155, 207n3. *See also* Kuki-Chin-Mizo ethnic group
Shira (film), 93

Simons, Chaim: as apologist for Goldstein massacre, 142–43, 206n22; memoir of first three years in Hebron, 64–66
Sinai Peninsula: Israel's return of, 79–80, 202n21; Yamit settlement (1982), 182, 202n21, 209n21
Smith, Jonathan Z., 126, 194n10
spatial practices and ideological uses of space, 5, 8–14, 23, 96–121, 194n6; beyond the settlement for economic transactions, 108–13; creating ethnic enclaves, 99, 102–5, 114, 200n2; to create place attachments, 8–12, 27–28, 103–4, 194n6, 194n9, 194n10; de Certeau on legibility of place and, 114; expansion of the private sphere through the *eruv*, 86–87; Israeli Independence Day celebrations in Hebron, 96, 99–102, 200n3, 200–201n6; in relation to ethnic exclusivism, 23, 102–5; resolute attachments to specific sites, 8, 11–12, 17, 194n8; settler incursions into Palestinian areas/property, 96–102, *98*, 107, 116, 190, 200n5; settler routes that expand boundaries, 97–99, 100–101; settlers' illegal entrance into Hebron (1968), 22, 50, 55–56; spatiality and power, 13–14, 96, 194–95n11, 200n1; stance of nonrecognition and indifference toward Palestinians, 20, 23, 97, 105–13, 209n10; that give rise to attachments to land, 47, 196–97n17, 207n30; and time/temporality, 8–9. *See also* borders/boundaries; origins, ideological settlers' emphasis on
Storm of Terror: A Hebron Mother's Diary (Leavitt), 87–89
Stump, Roger W., 99, 196n9
Supreme Court of Israel: and expedited conversions of Russian immigrants, 172; and Goldstein's grave and shrine, 149; Palestinians' cases brought to, 56; ruling on Beit HaShalom incident (2008), 176, 209n22; and settlers' disdain for secular law, 140
surveillance: security regime and Palestinians in Hebron, 42–45; settlers' distrust of state, 21, 144–45

Tamir, Shmuel, 60
Taub, Gadi, 197n11
Teḥiyah Party, 140–42
Temple Movements, 189–90
terrorism, 69, 116, 124, 144–45; Eyal (Kahanist group), 144; and Goldstein gravesite and shrine, 147–50, 207n31; Haetzni on, 140–42; Jewish Underground (ha-Maḥteret ha-Yehudit), 69, 140, 145, 199n12; Kach Party and, 69, 206n21; Palestinian attacks on settlers, 143; Palestinian suicide bombings, 18, 44; and Shalhevet Pas case, 90, 199n15; Yahya Ayyash, 116, 202n23. *See also* Goldstein massacre at the Tomb of the Patriarchs (1994); violence, settler
time/temporality, 8–9; halakhic division of laws to be implemented in human time/messianic era, 15–16, 117–18, 120; ideological settlers' interpretive strategies emphasizing continuous Jewish presence in Hebron (temporal linearity), 30, 34–35, 196n6; Nachshon's renewal of the Jewish cemetery as a historical cycle, 77–78; reiterative time, 8–9, 194n7; a soldier's retrospection, 41–45; temporal rotations and Jewish/Muslim religious holidays at Tomb of the Patriarchs, 137
Todorov, Tzvetan, 34, 196n5
Tomb of Rachel, 35
Tomb of the Patriarchs (Meʿarat ha-Makhpelah/al-Ḥaram al-Ibrahimi), 1, 3–4, 72–74, 126–50, 191, 193n2, 195n2; composite structures and Islamic character, 1, 127, 128, 193n2, 205n13; Djaouliyeh Mosque (Hall), 135; floor plan, *136*; Goldstein massacre (1994), 122–24, 132–50; government accommodations to settlers, 132–37; Hall of Abraham and Sarah, 135; Hall of Isaac and Rebecca, 123, 128, *129*, 135, 137, 138, 205n18; Hall of Jacob and Leah, 135; interior courtyard, 128–29; Jewish markers and inscriptions, 128–29, *129*; Jewish-Muslim aural competition and prayers, 132; Jewish-Muslim battle for control of, 3–4, 29–30, 128–29, 132–37, 204n8; Likud election rally in front of, *98*; maternalist activism at, 72–74; military presence, 1, 55, 122–24, 132, 134–35, 137, 205n14, 207n34; name designation(s), 195n2; Netanyahu's declaration (2010), 191; opened for Jewish visitation, 128; partitioning of the mosque, 4, 73, 74, 123, 127–28, 132–37, 150, 205n19; pluralistic character/concept of "religious pluralism" at, 132, 137, 205n15; pre-Goldstein massacre violence and confrontations at, 4, 132, 134–35,

Index

205n18; purification issues and graves, 130–32; remaking the sacred at, 1, 125–37, *129*; saint worship of Jewish patriarchs and matriarchs, 131–32, 205n12; secret circumcision of Nachshon's sons at, 72–74; security entrance to Jewish side, *133*; settler crowds and weekly Sabbath observance, 96–99, *98*, 107; and settlers' celebration of Israeli Independence Day, 100, 102; settlers' claiming entitlement to all of, 55–56, 100, 102, 104–5, 132–35; settlers' claims of Abraham purchase of Sarah's burial site, 3–4, 29–31, 36, 77, 135, 195n3, 205n19; settlers' disregard for secular law and regulations, 137–42; settlers' illegal entrance into Hebron to pray at (1968), 50, 55–56; temporal rotation to accommodate Jewish/Muslim religious holidays, 137; on twenty-first anniversary of Goldstein massacre, 207n34; UNESCO World Heritage site, 191; Yusufiyya Hall, 135
Toubi, Tawfik, 63
Tzarfati, Rabbi, *10*

UNESCO World Heritage sites, 191
Unit for the Coordination of Operations in the Territories, 52
United Nations Relief and Works Agency (UNRWA), 80

Vardi, Raphael, 135
violence, settler, 23–24, 69, 122–50, 186–88; and absence of fixed borders, 62–63, 197n8; alliances between settlers and the military, 122–24, 132, 200n4, 205n14; cognitive dissonance and, 124–25; complicit attitudes toward, 146–47; and concept of religious pluralism, 132, 205n15; disregard for secular law, 137–42; distrust of the state and government surveillance, 21, 144–45; earliest settlers, 62–63, 69; emergency security arrangements, 123; everyday practices, ideologies, and local conditions, 24, 125, 187; far-right apologists, 140–45, 150; fieldwork dilemmas, 186–88; Goldstein massacre, 122–24, 132–50; gun possession and armed settlers, 24, 51, 96–97, 142, 143, 187–88; Haetzni and, 140–42; harassment of Palestinians, 45–46; and Israeli government, 149; justifications exploiting sense of threat and vulnerability, 142–45, 150, 206n24, 206n25; Kach Party, 69, 140, 198n12, 206n21; memorialization of, 145–50, 207n31, 207n32; post-Oslo Accords, 124–25; and Rabin assassination, 144–45, 206n26; settlers' views of, 45–46; at shared sites, 130, 204n10; and symbolic biblical violence, 203–4n6, 207n31; and terrorism, 116, 124, 144–45; women's takeover of the Daboya clinic, 84. *See also* Goldstein massacre at the Tomb of the Patriarchs (1994)

Waldman, Rabbi Eliezer, 54, 66
Weil, Shalva, 207n5
Weiss, Hadas, 208n16
Weizmann, Ezer, 83
West Bank, 12, 35; biblical terms used for, 12, 35; Judea and Samaria, 12, 35, 199n16; and 1967 war, 6, 33; population growth and fertility rates, 199n16; post-Oslo spatial reorganization of, 40–41, 43–44. *See also* Occupation of the West Bank
Western Wall (Jerusalem), 189, 205n17
Wolfe, Patrick, 7
Women in Green, 85
women settlers: Gaza, *181*; and gendered aspects of Jewish observance, 70–71, 85, 91; large families and women's status, 91–92; prolific reproduction, 71, 85, 89, 91–93, 94, 199n18, 209n19; religious education, 85; scholars on women in fundamentalist communities, 71, 84–85; Women in Green, 85. *See also* maternalist activism and property takeover in Old City of Hebron; Nachshon, Sarah
World Zionist Organization (WZO), 67

Yadin, Yigal, 83
Yamit settlement in the Sinai, dismantling of (1982), 182, 202n21, 209n21
Yemeni Jews, 110
Yiftachel, Oren, 102–3
Yishuv (Jewish settlement of Palestine, 1882–1948): demographics and security, 94; "frontierism" of pre-state settlements, 108
Yochai, Rav Shimon bar, 48
Yom Kippur War (1973), 66

Zar, Moshe, 199n12
Ze'evi, Rehavam, 196n14, 202n22
Zertal, Idith, 55, 57–58, 127

Zionism/Zionists, 35, 39, 121; as a comparatively secular nationalism, 118, 203n28; demographic emphasis, 94, 208n8; Drobles's references to settlers over the Green Line, 68; and erasure of the Palestinian presence, 39; ideological settlers' claims of continuity with, 68, 197n11; Paine on Zionism and the religious certainty of settlers, 103–4, 201n8; and place-based aspect of settling, 194n9; pre-state settlements and land ownership, 39, 94, 108, 201n13

Zo, 155, 207n3. *See also* Kuki-Chin-Mizo ethnic group

Zohar, 48

Acknowledgments

This book has been researched, written, and revised over a long period of time. At the University of Chicago, the people who directly commented on an early version of the manuscript include the late Paul Friedrich, Jim Fernandez, Rashid Khalidi, and Arjun Appadurai. Paul Friedrich sharpened my understanding of the convergences between ideology, land, violence, and language that are analyzed in this book. Jim Fernandez introduced me to the perils and pleasures of writing ethnography, and Rashid Khalidi encouraged me to explore both Palestinian and Middle Eastern history in great depth. Arjun Appadurai greatly added to my engagements with the comparative dimensions of ethnic conflict, colonialism, and globalization. In addition, other important influences included Moishe Postone on Marx, Leora Auslander on gender, and Jonathan Z. Smith on the history of religion. Colleagues from that period who have continued to inspire me with their scholarship include Engseng Ho, Nadia Abu El-Haj, Paul Liffman, and Mala de Alwis. The late Tania Forte was a key interlocutor on Israeli, Palestinian, and Middle Eastern issues. The early stages of fieldwork and writing were financed by grants from Fulbright-Hays, CASPIC/John T. and Catherine D. MacArthur Foundation, and Harry Frank Guggenheim Foundation. During a year in New York, I also greatly benefited from the insights of Samira Haj, Talal Asad, and Michael Taussig.

At Bryn Mawr College, I learned a good deal from my students, including Sarah Alibabaie, Sara O'Connor, Diana Tung, Kali Noble, Dinu Ahmed, and Hind Eideh, and I am grateful for their insights. Bryn Mawr College's Faculty Research Funds allowed me to conduct important follow-up fieldwork. Colleagues at Bryn Mawr who directly contributed to the book include Debbie Harrold, Marc Ross, Rick Davis, Azade Seyhan, Sharon Ullman, Peter Magee, Kalala Ngalamulume, David Rabeeya, Camelia Suleiman, Rick McCauley, Michael Allen, and Kristine Koggel. Annette Baertschi and Richard Stahnke cheered me on to the home stretch. Naomi Koltun-Fromm, at

Haverford College, provided valuable comments on religion and violence, as did Farha Ghannem, at Swarthmore College, on West Bank settlement. At Lafayette College, Andrea Smith contributed to my understanding of the comparative dimensions of settler colonialism. During earlier research trips to Israel/Palestine, the Truman Institute at Hebrew University and the Van Leer Institute extended valuable institutional affiliations.

I am also grateful to Israeli, Palestinian and international NGOs for facilitating this research, particularly B'Tselem, Breaking the Silence, al-Haq, Passia, CPT Palestine, and Muwatin. Several key informants living in Kiryat Arba and other West Bank settlements took the time to explain their positions and aspirations to me as an outside researcher. I thank them for their engagement, without which I would not have understood the history and legacy of ideological settlement with as much depth. Finally, I am indebted to the late Michael Feige and the late Ehud Sprinzak, as well as Stephanie Genkin, Ramsey I. Jamil, May Jayyusi, Lena Jayyusi, Ifat Maoz, Gadi Taub, and Idith Zertal for their help and valuable insights on Israel, Palestine, and settlement history.

As a Regional Faculty Fellow at the Penn Humanities Forum, I benefited from the input of the "Adaptations" working group and in particular from the comments of Jim English, David Grazian, Warren Breckman, Sally Scholz, and especially Ayako Kano, whose readings and resolve contributed to my finishing the book. I am also thankful to Laura Helper-Ferris for her editorial insights, Peter Wissoker for further comments, and Smadar Shtul for help with Hebrew transliteration. Peter Agree at the University of Pennsylvania Press provided encouragement as well as expert advice and made this book a reality. I am also indebted to Managing Editor Erica Ginsburg for her editorial insights and Jennifer Shenk for her meticulous copyediting. Two anonymous reviewers provided extensive and valuable comments on the entire manuscript that helped me revise it. I am also grateful to Kaushik Ghosh and Sarada Balagopalan, Karin Badt, Ruth Chojnacki, Zita Laos, and Carol Scherer for their intellectual engagement and long-standing friendship. Finally, I want to acknowledge my father, Jaime León Neuman for his generosity and wit, and my late mother, Dorothea Wohl Neuman, for being an important early teacher. As with any work of scholarship, the book's analyses and conclusions are my own and do not necessarily reflect the views of those mentioned above.

Parts of the following three chapters were previously published and appear here in revised form: Chapter 3, "Motherhood and Property Takeover," first appeared as "Maternal 'Anti-Politics' in the Formation of Hebron's Jewish Enclave," *Journal of Palestine Studies* 33, no. 2 (Winter 2004); Chapter 5, "Religious Violence," was published as "Religious Nationalism, Violence and the Israeli State: Accommodation and Conflict in the Case of the Religious Settlement of Kiryat Arba," in *Religion und Nation, Nation und Religion: Beiträge zu einer unbewältigten Geschichte*, edited by Michael Geyer and Hartmut Lehmann (Göttingen: Wallstein-Verlag, 2004). A section of Chapter 6, "Lost Tribes and the Quest for Origins," was published as "Alienations of Exilic Return: Russian Immigration and 'Ingathering' in Hebron," in *Realms of Exile: Nomadism, Diasporas, and Eastern European Voices*, edited by Domnica Radulescu (Lanham, MD: Lexington Books, 2001). Last but not least, I am deeply grateful to Matthew J. Hill, my spouse and dear friend, who has read far too many drafts and lived through the many ups and downs of this long and difficult research project. It is to him that I dedicate this book.